TAIWAN

INDON
SOUTH

*PACIFIC
OCEAN*

20

0°

do
TERNATE
TIDORE
*OLUCCA
SEA*

HALMAHERA
(JAILOLO)

24

MOLUCCAS

SERAM
• Ambon

IRIAN JAYA

PAPUA
NEW
GUINEA

BANDA
ISLANDS

BANDA SEA

-4 J

fron

• Dili

TIMOR

TIMOR SEA

AUSTRALIA

IMAGES OF ASIA

Spices

Titles in the series

Spices

The Story of Indonesia's Spice Trade

JOANNA HALL BRIERLEY

KUALA LUMPUR
OXFORD UNIVERSITY PRESS
OXFORD SINGAPORE NEW YORK
1994

380. 141383   9/94 TCF

MBN-L3 i/00

Oxford University Press

Oxford New York Toronto
Delhi Bombay Calcutta Madras Karachi
Kuala Lumpur Singapore Hong Kong Tokyo
Nairobi Dar es Salaam Cape Town
Melbourne Auckland Madrid

and associated companies in
Berlin Ibadan

Oxford is a trade mark of Oxford University Press

Published in the United States
by Oxford University Press, New York

© Oxford University Press 1994
First published 1994

British Library Cataloguing in Publication Data
Data available

Library of Congress Cataloging-in-Publication Data

Brierley, Joanna Hall, 1943–
Spices: the story of Indonesia's spice trade/Joanna Hall Brierley.
    p.   cm.— (Images of Asia)
Includes bibliographical references and index.
ISBN 967 65 3039 5
1. Spice trade—Indonesia—History. I. Title. II. Series.
HD9210.I62B74 1994
380.1'41383'09598—dc20
93–36554
CIP

Printed by Kyodo Printing Co. (S) Pte. Ltd., Singapore
Published by Oxford University Press,
19–25, Jalan Kuchai Lama, 58200 Kuala Lumpur, Malaysia

For Dumbo and Neil

Preface

WHEN I originally broached the idea of a book on the spice trade of Indonesia, I did not visualize the depth of the topic nor the fascinating detail that was to be uncovered. Working on the book has primarily entailed a study of the history of the archipelago, more particularly the foreign contacts who contributed their share of culture to the time-honoured traditions and customs of the islands that make up the Indonesian archipelago. The study has been botanical also, involving a close look at those spice plants which were the very reason for world voyages of exploration and discovery. Spices began the commerce between East and West that still exists today. The final chapter of the book examines the role of spices in the culture of this richly artistic country, where they are used to dye cloth and inspire decoration on batik and silver, and where they form an integral part of traditional ceremonies and festivities.

I thank my mother, fascinated with the Spice Islands, for the original idea for this book, and also my husband Neil for his help with illustrations, and interest in the research.

I am most grateful to the Picture Library and to Philippa Glanville of the Victoria and Albert Museum, London, who provided her own picture of a seventeenth-century nutmeg grater, to the Royal Botanic Gardens in Edinburgh for permission to use their illustrations, to Antiques of the Orient in Singapore for the use of their old prints, and to Dra Suwati Kartiwa and Ibu Paulina Suitela of Jakarta's National Museum for allowing me to photograph from the Museum's collection.

Wendy Veevers-Carter, John Guy, Grace Inpam Selvanayagam, and the National Museum of Singapore kindly allowed me to use photographs from their collections. For providing and taking pictures, I thank Kirsten Neiser-Schlaudraff, Hilde Ardie, Barbara Johnson, and Mochtar Lubis.

My appreciation for the use of their libraries and for information on the spice trade goes to Dr E. Edwards McKinnon, Frank Mummery, the Herbarium Bogoriense and the Research Institute for Spices and Medicinal Crops in Bogor, Ian Neil, Philip Hausel, and Roel van den Brink of International Flavors and Fragrances, the family-run Darmi Jamu factory of Jakarta, *kretek* producers PT Djarum and PT Gudang Garam, and, finally, the Ganesha Volunteers for their library and resources.

Sacrificing the romance of old names, I have used contemporary spelling in the text which I hope will be less confusing to the reader.

Jakarta   JOANNA HALL BRIERLEY
April 1993

Contents

1
Spices in Antiquity

WHAT is a spice? The term is derived from the Latin *species*, meaning 'an article of special value'. The *Concise Oxford Dictionary* defines a spice as an 'aromatic or pungent vegetable substance ...'. The ancient functions of spices were as ingredients in perfumes and cosmetics, fumigants and medicine, and as condiments. In ancient Egypt, spices were employed in the embalming of bodies; they were the myrrh and frankincense of Buddhist China. The Roman Apicius the Epicure wrote the first cookery book in 1500 BC during the reign of Augustus, using pepper, whole or ground, in every recipe. Dioscorides composed the first medical manual, *Materia Medica*, in AD 65, in which he wrote on the contribution of spices to pharmacology. Many spices remain today official drugs. In the Europe of the Middle Ages, spices, especially pepper, disguised the flavour of heavily salted meat preserved in the autumn to last till the following spring.

The story of the trade in eastern spices precedes written history. It began with the story of world exploration, from the West to the tropical islands of the East, in search of spice. To the indigenous local populations in the East, spices have been known for thousands of years, being used to add taste to an insipid carbohydrate diet and to make traditional medicines. China traded in Indonesian cloves from the third century BC, and in ginger, which was widely used; the sage Confucius (551–479 BC) was apparently never without it during his meals. In the second century BC, spices from the East Indies, what is today Indonesia, were brought by sea and land to China, and were among the cargoes carried by camel along the Silk Road to the West. The perfumed woods—cinnamon and cassia—were carried in outrigger canoes to the East African coast in the first century AD. It was in this century, too, that the Romans, who were extravagant users of spices, attempted to break the Arab transport monopoly by utilizing the newly discovered monsoon winds in an effort to open up their own trade with India (Purseglove et al., 1981).

The transport of spices by land and sea was for centuries a monopoly of Arab middlemen. One of their ancient trade routes originated on India's Malabar coast, where nutmeg, mace, and cloves, brought from the Moluccas, were carried to the Persian Gulf and along the Euphrates Valley to Babylon. Recent archaeological excavations in this valley, the Mesopotamia of old, have revealed charred clove remains dating to 3000 BC (*Past Worlds*, 1988). Another sea route used by the Arab traders went from India, round the coast of Arabia, and through the Red Sea to Egypt—to satisfy the requirements of Egyptian importers—then across the Mediterranean where spices were bought by the ancient Romans and Greeks. It was the Greek city of Tyre that was an important commercial centre between East and West until conquered by Alexander the Great in 332 BC. In that same year, he founded Alexandria, which then became the trading centre for spices between the Orient and the Mediterranean.

Traded spices were as highly valued as gold and jewels: the Queen of Sheba brought precious stones, gold, and spices to Solomon in 992 BC, and 3,000 pounds of pepper were demanded by the invading Alaric, King of the Goths, to withhold the sacking of Rome in AD 408. When the city fell two years later, Constantinople became the eastern capital of the Roman Empire, a city around which important overland and sea trade developed with the East. Basra, at the head of the Persian Gulf, became an important trading centre for spices. It was here that physicians first mixed spices to make tonics and medicines which were later to be copied by the apothecaries of Europe.

During the Crusades (AD 1096–1291), Venice became the spice and trading centre of the Mediterranean. The enormous profits still being made by the city in the fourteenth century were to help finance the Renaissance. When Constantinople fell to the Turks in 1453, overland trade was impeded and it became increasingly imperative to find a sea route to the Indies and the source of spice. This was achieved in the sixteenth century by the Portuguese, who broke the Venetian monopoly.

J. Innes Miller (1969), writing on the ancient history of spices, informs us that 'The art of their various uses was common among civilized peoples, in their homes, their temples, their public ceremonial, and in the seasoning of their food and wine. A peculiar attribute was their medicinal power. That they were dried and of small bulk made them easy of transport, and their rarity a form of royal treasure.'

2

Sources of Spice

THE East Indies, the source of so many traded spices, are strategically located between mainland Asia and India on the East–West trade route. Acting both as an entrepôt state and as an active contributor to trade, the Indies has attracted foreigners for millenia. The largest archipelago in the world, the 13,000 islands which now form the Republic of Indonesia are strung across 5000 kilometres of sea.

The archipelago had two distinct centres of spice. First was the clove- and nutmeg-producing Moluccas to the east, which formed a volcanic mini-archipelago of around 1,000 islands. Only a handful of these were spice-producing. They are situated north of the Equator in an area known geographically as Wallacea, after Alfred Russel Wallace, a contemporary of Charles Darwin, who spent years in the area documenting the flora and fauna. (The results were published as *The Malay Archipelago*, which remains a classic reference on the region.) Before the arrival of the Portuguese in the sixteenth century, little was written or known about the Moluccan islanders, who had been trading their spices for about 2,000 years before the arrival of the first Europeans.

Cloves were native to only five islands in the Moluccas: the rival sultanates of Ternate and Tidore as well as Motir, Makian (Machian), and Batjan (Bachian), all lying off the western coast of the giant Halmahera (once known as Gilolo) in the northern Moluccas (Plate 1). These were the original Moluccas. Later, the name also included the nutmeg and mace islands of Banda to the south, of which Neira, Lontar, and Ai were the centres (Plate 2).

The early interisland trade in Moluccan spices was linked to the primary food source of the islands, the sago palm *Metroxylon sagu*. This enormously productive palm provided essential food for the small, volcanic, inner islands of the Moluccan archipelago on which spices and coconuts were all that could grow. These inner islands profitably traded their spices for the sago grown on the outer islands of Halmahera, Seram, Kei, and Aru (Ellen, 1979). The Portuguese Jesuit Antonio Galvão was to remark in 1544 that clove, nutmeg, and sago were 'the most useful trees of these islands', while Wallace wrote that freshly baked sago cakes became quite a delicacy with the addition of grated coconut.

By the twelfth century, the Moluccas were established as a substantial trading centre. The powerful sultanates of Tidore, and more especially Ternate, wielded considerable political power over the surrounding islands, whom they forced to pay tribute, often in the form of slaves or foodstuffs. The lucrative sale of spices was very much in the control of the nobility: they supervised the spice plantations, weighed out the spices in their own homes, and received part of each crop as payment. Ternate's governorship of the large island of Seram continued until 1643, and she also controlled parts of Ambon when the Dutch established a presence on the island in the seventeenth century.

Moluccan trade expanded in the fourteenth century with an increased European demand for spices. It was dominated by Javanese merchants, who not only recognized the economic significance of these spices but, importantly, saw the Moluccas as a ready market for Java's rice. Javanese, Malay, and, to a smaller extent, Chinese traders marketed their wares in the islands' tiny market-places, beaches, and streets. As well as the much needed rice from Java, these merchants brought into the trade Chinese porcelain, coins, brass gongs, and cloth, and Indian cotton, silk, and beads, which were bartered against the local spices, tree resins, bird of paradise plumes, and sea slugs.

The Moluccans themselves were not a seafaring people; it was only the Bandanese who used their fleets of oar-propelled *kora-kora* canoes to ply small trade among the islands (Plate 3). On a larger scale, along with the Javanese and Asian middlemen, local island trade throughout the archipelago was carried out by the legendary Bugis seafarers, the sea gypsies of Indonesia, who originated from the coasts of Sulawesi and were the traders and pirates of the eastern seas from early days. They plundered

5

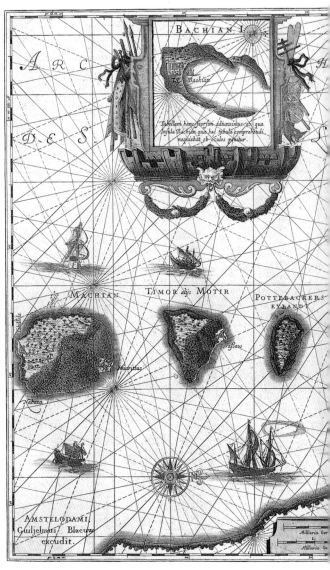

1. The five clove-producing islands of the northern Moluccas, by W. J. Blaeu, *c.*1650, from R.T. Fell, *Early Maps of South-East Asia*, Singapore, Oxford University Press, 1988.

MOLVCCÆ
INSVLÆ
CELEBERRIMÆ

LAGO

ORO

Hærij

TERNATE

Grammalamma

Oleoge dort

MITIARA

Maleyo

Bay van
Gilolo

TIDORO

GILOLO I.

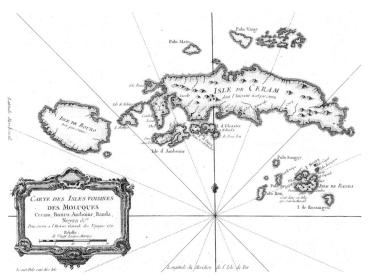

2. The nutmeg-producing Banda Islands, south-east of Ambon, from Antoine François Prévost, *Histoire Générale des Voyages*, Paris, 1946–61. (Antiques of the Orient, Singapore.)

ships—and thus were to be a continuous source of harassment to later European sea trade—transported slaves, and utilized the monsoons to collect cargoes of forest and sea produce from the Moluccas. These they traded with the Sumatran kingdom of Srivijaya and later with ports in Java. Bugis *pinisi* in modern times still ply their interisland trade throughout Indonesia, and can be seen unloading their cargoes in Java's port of Sunda Kelapa on the outskirts of present-day Jakarta. On the eve of European arrival in the sixteenth century, this was a port described by the Portuguese Tomé Pires as 'magnificent … the most important and best of all. This [port] is where the trade is greatest and whither they all sail … the merchandise from the whole kingdom comes here' (Cortesão, 1944).

The second distinct source of spice was the region comprising the islands of Java and Sumatra in the west of the archipelago (see front endpaper). Though they lacked the romantic associations of the spice islands to the east, Sumatra and Java were every bit as

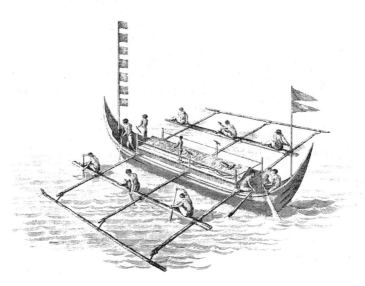

3. *Kora-kora*, from Thomas Forrest, *A Voyage to New Guinea and the Moluccas, 1774–1776*, London, 1780.

influential in luring foreign trade. It was from here that the highly valued pepper came: of all the spices of Indonesia it was the greatest in demand and reaped the highest profits. From these islands, too, came the perfumed resins and the aromatic crystals of camphor which had been traded to China since the time of Christ.

In the seventh century, the powerful maritime kingdom of Srivijaya, centred on Sumatra's east coast at what is now Palembang, dominated the sea routes from East to West. This early trading state, which controlled both the Malacca and Sunda Straits, exerted a strong influence over neighbouring Java, the Malay Peninsula, and ports along the Bay of Bengal. Indonesian shipping was noted there in the first century AD by a Greek sea captain in his manual on Indian Ocean trade, the *Periplus of the Erythraean Sea*.

Much of the prosperity of the early trading kingdom of Srivi-jaya, which remained influential until the fourteenth century, was due to Chinese demand for the luxury items of Sumatra's forested hinterland. Srivijaya was an assiduous mission-sender to China, whose own role in early seaborne trade was relatively minor. It was not until the outward-looking Tang Dynasty of the eighth century that China was to conduct a vigorous mari-time trade with Indonesia, which was to be maintained during the following Song Dynasty of the tenth century.

3
Early Trade Routes

IN the first century AD, a Greek sea captain, Hippalus, discovered the monsoons, seasonal winds which were to control shipping and trade before the advent of steam. These winds blow in alternate directions over the same route at half-yearly intervals. The north-east monsoon, which blew over South-East Asia from December to March, took early seaborne traders to the West. The south-west winds of June till September blew them home, and carried those seeking spices to the East. In the lull between winds, ships sought harbour and their captains had time to trade or effect repairs to their vessels.

The use of the monsoons was to reduce the importance of the overland carrying route and to open up increased sea trade with India. It is noted by the anonymous author of the *Periplus* in 1 AD that ports on India's Malabar coast were now being frequented by merchants other than Arabs, who had previously dominated the spice trade and probably made use of these facilitating winds even before Hippalus's discovery. With the round trip from the Mediterranean to India now taking no more than a year, spices became more accessible and less of a luxury. Their prices were reduced, and some of the mystery surrounding the source of spice, and its trade, carefully perpetuated by the Arabs, was lifted.

The Cinnamon Route

Employing these north-east monsoon winds, Indonesian sailors in outrigger canoes carried cargoes of cinnamon, cassia, and probably other spices to Madagascar and the east coast of Africa in the first century AD. The Roman encyclopaedist Pliny wrote an account of the trade (quoted in J. I. Miller, 1969) recording admiration for these intrepid seamen: 'They bring [spices] over vast seas on rafts which have no rudders to steer them or oars to push … or sails or other aids to navigation … but instead only the spirit of man and human courage…. These winds drive them on a straight course, and from gulf to gulf. Now cinnamon is the chief object of their journey, and they say that these merchant sailors take almost five years before they return, and that many perish.' Pliny was rather dismissive of the return cargo of glass, bronze, clothing, and jewellery as being dependent on 'women's fidelity to fashion'.

The influence of the spice traders from Indonesia can be discerned in Madagascar today in the cultivation of dry and wet rice and the ceremonies associated with its planting and harvesting, the cattle that work the plough, the types of fish-traps used on the island, the houses built on stilts, and the many words in the Malagasy language which have Indonesian roots. Indeed, one seventeenth-century British traveller to Madagascar, Richard Boothby, was so confused by the Eastern character of the place that he described it as 'in Asia, in the vicinity of East India' (McBain, 1988).

It is also thought that one of coastal East Africa's traditional sailing vessels, the *mtepe*, a type of dhow no longer in use, owed its origins to the pegged and sewn outrigger craft of the Indonesian islands, which use no nails in their construction (Plate 4).

After unloading part of the cargo in Madagascar, the outriggers sailed on to Rhapta (Map 1). Although nothing remains today of this important settlement, referred to in the *Periplus* as 'the very last market-town of the continent', it was thought to have been situated in the Rufiji Delta, south of today's Dar es Salaam, and was probably the southerly headquarters for Arab influence in the region.

From Rhapta spices were carried north to the arid coast of Somalia, which was known in antiquity as Punt, or the Land of Incense, because it was believed that cinnamon grew there. The Horn of Africa, on the tip of Cape Guardafui in Somalia, was the Cape of Spices, known as the Aromaton Emporion or the ancient Mart of Spices. There were specialized entrepôt ports here where enormous quantities of Indonesian aromatic gums were graded and weighed before being transported further to Egypt and the

4. An Indonesian outrigger depicted on a stone relief at the temple of Borobudur, Central Java. (John Guy.)

Mediterranean. Roman coins and glass, Nubian pottery, and Chinese ceramic sherds found in the area are an indication of early international trade (Chittick, 1979).

As mentioned earlier, this lucrative carrying trade was in the hands of Arabs who kept secret the Eastern source of the spices. They frightened off other would-be traders with invented tales of giant birds whose nests were made of cinnamon twigs, collected at great risk from swamps guarded by ferocious bats and giant snakes (J. I. Miller, 1969).

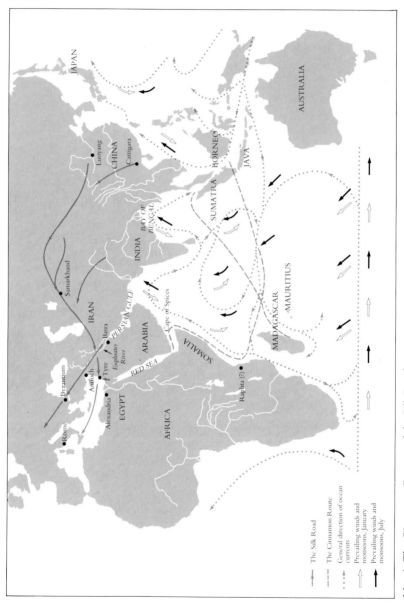

Map 1. The Cinnamon Route and the Silk Road.

The Silk Road

At the same time as Indonesians were plying the Cinnamon Route to Africa, they were trading with India and China, whose overland and maritime trade routes had long linked their countries to the West. Indeed, caravans had carried spices to the West from biblical times: in the Book of Genesis, Joseph is sold by his brothers to the Ishmeelites, whose camels were bearing 'spicery and balm and myrrh, going to carry it down to Egypt'.

The Silk Road, which crossed the steppes and deserts of Central Asia, was opened in 206 BC during the Chinese Han Dynasty. It owed its name to the trade in silk, woven in China from around 2700 BC and sent westwards as a placatory gift to nomadic border tribes. Later, silk became fashionable and much sought after in the Roman Empire. Fragments of silk have been found as far afield as Athens, the Ukraine, and the British Isles, testimony to the extent of the trade in this product (*Past Worlds*, 1988). Traded along with silk were furs, lacquer, jade, bronze, and, importantly, spices.

The Silk Road was not a single path from East to West, but, rather, several caravan trails between one trading centre and another. Some routes followed both land and sea and utilized available rivers such as the Tigris and the Euphrates. Trade trails were dictated by the various producing and collecting centres. These were punctuated by offshoots and diversions to areas of commercial interest. Because bandits were a hazard, watch-towers and forts were erected along the road and attempts made to police the trails. Cities offering a safe haven flourished as trading stations, and in the caravanserais built along the route, rest was offered to weary travellers. Here they could renew their mounts and trade their goods, which were to change hands many times before they reached their final destination, becoming more expensive with every transaction.

The Eastern terminus of the Silk Road was Luoyang, capital of the Han court. There was a more southerly terminus at the port of Cattigara, however, which, because of its location on the Gulf of Tonkin, had sea links with the Indonesian Moluccas.

15

From Cattigara, Moluccan spices travelled by a combination of land and sea routes to their Western destinations, or were carried overland from the port to Luoyang, where the caravans assembled and the merchants met and loaded their beasts of burden: Bactrian camels, horses, elephants, or yaks. Travelling overland to such western termini as Antioch and Tyre, cargoes would then go on by sea to the spice markets of Alexandria and Rome where, in AD 1, the Emperor Tiberius was grumbling at the cost to the Roman economy of 'exotic Asian products'.

The volume of trade varied as political fortunes rose and fell, although it was particularly large during China's Tang Dynasty (AD 618–906). Commercial contact by the land route was maintained between China and the West until the thirteenth century, when seaways—and the monsoon winds—were increasingly utilized. Even so, the Silk Road continued to function, though to a lesser degree, until the European trading companies of Britain and Holland, with their superior sea power, dealt the land trade a fatal blow.

4
Spices of Trade

WITH a weather pattern unchanged for centuries, the climate of the East Indies is universally hot and moist throughout the year. This, combined with regular volcanic activity, produced the fertile soil that nineteenth-century naturalist Alfred Russel Wallace remarked 'teems with natural productions which are elsewhere unknown. The richest of fruits and the most precious of spices are here indigenous.' Prominent among these were the spices of trade: cloves, nutmeg and mace, pepper, and perfumed woods and resins. Less well known, but traded early to the West also, were the spicy rhizomes of the ginger (*Zingiberacae*) family: ginger, turmeric, galangal, and the pungent seeds of cardamom.

Cloves Eugenia caryophyllus Cengkeh

Along with the southerly nutmeg islands of Banda, the five clove-producing islands of the northern Moluccas formed the original Spice Islands of history—the Moluccas.

The Indonesian term for cloves is *cengkeh*, which is thought to originate from the Chinese *tkeng-his*, meaning 'scented nails'. The spice's resemblance to nails is noted in other languages also: cloves are *clou* in French and *nagel* in Dutch.

The commerce in cloves is an ancient one; the clove was the earliest traded spice. The Chinese, trading with the northern Moluccas during the Han Dynasty (206 BC–AD 220), had a court rule that cloves were to be sucked to sweeten the breath before speaking to the Emperor. Cloves were used in both India and China as ingredients in perfumes, medicines, and as flavourings for food (J. I. Miller, 1969), and the Indians, as mentioned in the epic, the *Ramayana*, of 200 BC, used cloves to aid digestion and to fasten the betel-pepper leaf around the areca nut that is chewed throughout South-East Asia. Early customs documents record cloves among the cargo of spices that travelled to East Africa by outrigger canoes in the first century AD, from whence Arab merchants carried them to Egypt. Cloves were also known in the Mediterranean by the fourth century, where it is documented that 150 pounds were presented to the Bishop of Rome, in AD 314, along with other precious commodities such as gold, silver, and incense. By the eighth century, cloves were in use throughout Europe, utilized with other spices to flavour and preserve food, and to sweeten the air (Purseglove et al., 1981).

Marco Polo was the first European to give an account of his travels in the East, and he wrote, in the thirteenth century, of cloves growing on islands in the China Sea. They were recorded also by the diarist on Ferdinand Magellan's voyage to the Spice Islands in 1519: significantly, the sole surviving captain of the expedition was rewarded for his cargo of spices with a coat of arms that depicted twelve cloves, three nutmeg, and a stick of cinnamon. Francis Drake found cloves in Ternate on his voyage round the world in 1577. He took on such a heavy cargo that he had to lighten his load by jettisoning 3 tons to refloat the *Golden Hind* (Purseglove et al., 1981).

In their sixteenth-century attempts to monopolize the trade in Indonesian spices, the Portuguese established a trading post on clove-producing Ternate. Trade was based entirely on barter, with Ternate exchanging her cloves for Indian cloth carried by the Portuguese from their possessions on India's Malabar coast, for the porcelain and ivory of the Chinese merchants, and the much needed rice from Java, whose merchants re-traded the cloves in their ports.

The Dutch ousted the Portuguese in the seventeenth century, and shifted the centre of clove cultivation from the northern Moluccas to Ambon, south of Ternate. This was to be the Dutch headquarters for the Moluccas, from which base they severely restricted the production of cloves in the area to ensure the highest possible return on their monopoly. For the Hollanders, it was to be a far more successful monopoly than that achieved by the Portuguese, under whom a degree of local trading had continued. But for the spice islanders, forced to sell their now limited

5. Clove tree, from Johan Nieuhof, *Voyages and Travels to the East Indies, 1653–1670*, Singapore, Oxford University Press, 1988; reprinted from the second part of *A Collection of Voyages and Travels ...*, London, A. & J. Churchill, 1704.

production of cloves to the Dutch at a fixed price, and forbidden to trade themselves, it meant impoverishment as their source of income was removed. This Dutch monopoly was broken in 1770 when clove seedlings were smuggled to Mauritius, and from there to Zanzibar.

Cloves grow on a laurel that is the most aromatic member of the Myrtle (*Myrtaceae*) family (Plate 5). W. Veevers-Carter, in *Riches of the Rain Forest* (1984), informs us that 'young clove leaves stand up perkily, light green and shiny except, in the newest growth, when they are pink or red ... clove trees are particularly easy to recognize in that they "flush" continuously throughout the year, a pretty sight in the hillside plantations'.

The clove of commerce is the unopened flower bud which is picked just before it is ripe and then dried in the sun to seal its fragrant properties. It is then ready for sale without further processing, its quality assessed on appearance. Good quality cloves are plump and light brown in colour.

From about four years old, trees bear a small amount of fruit, but it takes another twenty years for the clove to produce its best yields, with a bumper crop produced every three to seven years thereafter. Trees remain productive for many years, with a mature tree producing around 15 kilograms of cloves a season. It is not so surprising perhaps that a giant tree in Ternate town is reputed to be 400 years old and is still yielding a good harvest.

Harvesting is carried out from June to December, in a pattern which has hardly changed over the years. Cloves are hand-picked when the buds are pink, and are collected in clusters from the lower branches while standing on the ground. Trees have to be climbed to collect most of the crop, and this is done carrying a basket and rope. The cloves, with their leaves and twigs intact, are picked and put in the basket which is lowered to the ground when full. Separating the twigs and leaves from the cloves is a family affair, and then the fresh cloves are dried on trays in the sun. No longer exclusive to the Moluccas, though these islands remain the largest supplier, cloves grow also in East Java, South and West Sumatra, and in Sulawesi. Trays of drying cloves, with their pervasive fragrance, are a common sight in Indonesian villages during the harvesting season.

Approximately 150 000 tonnes of cloves are produced annually, but only a fraction are exported, for instance, 360 tonnes in 1991. Most of the clove crop is consumed by the local cigarette industry, in which shredded tobacco is mixed with cloves to produce the highly aromatic *kretek* cigarette.

Cloves are utilized either whole, ground, or in the form of an oil called eugenol, which is distilled from the bud (which produces the finest quality oil), the stem, or the leaves of the tree.

The essential oils of spices have been extracted by man for almost fifteen centuries. Warm, spicy clove oil is employed pharmaceutically for its excellent disinfectant qualities, in dentistry for its analgesic properties, and in some medications to disguise the unpleasant taste. In the East, cloves are used in herbal medicine and as a flavouring ingredient in the areca nut chew.

Fifty per cent of all fine perfumes employ the fragrance of cloves. At the lower end of the market, cloves scent soaps and detergents and season commercially prepared foods such as sauces, pickles, and chutneys.

Nutmeg and Mace Myristica fragrans Pala

Nutmeg and mace are indigenous to the Moluccan Bandas, tiny islands at the southernmost tip of the Spice Islands. Nutmeg arrived in Europe after the clove, the first record of it being from Constantinople in AD 540. It was probably traded by the Arabs via India from where its Sanskritic name, *pala*, comes (Purseglove et al., 1981). *Pala* is also the name given to this spice in Indonesia. Mace is known as *bunga pala*, or nutmeg flower.

Nutmeg and mace are unique in that the two distinctly different spices grow together on the one tree, a glossy-leaved evergreen which can grow as tall as 20 metres (Colour Plate 1). The fruit is the size and colour of a peach, bursting open when ripe to reveal the crimson mace that encircles a shiny black nut. Inside this nut is the brown nutmeg of commerce, while the mace is the edible aril, or coating, of the seed (Plate 6).

By the twelfth century, nutmeg and mace were in wide use in Europe, both in kitchens where they were used in large quantities in sweet dishes, and in apothecaries. In fourteenth-century Germany, one pound of nutmeg cost as much as seven fat oxen (Rosengarten, 1969). Mace, however, was the more expensive spice, and the most in demand. Not appreciating the fact that the two spices grew together, the Dutch and British constantly requested their colonies to grow more mace and less nutmeg!

Initially growing wild throughout the Banda Islands, under Dutch colonial rule nutmeg production was centred on the islands

6. Freshly picked nutmeg, the mace encircling the nut. (Neil Brierley.)

of Neira, Lontar, and Ai. The Dutch wanted a monopoly on the sale of nutmeg and mace, as they had with cloves, and in 1621 they brutally subjugated the Bandas in an effort to enforce this. The Bandanese were forced to sell all their spice crops to the Dutch at a fixed price and made to buy unwanted imported goods in return. This policy was as economically disastrous for the Bandanese as it had been for the clove-growing islands to their north.

An unsuccessful attempt to break the Dutch monopoly was made in 1818 when nutmeg was introduced to Zanzibar, in East Africa, along with cloves. They never thrived. Sir Thomas Stamford Raffles, however, grew them with more success during his governorship of the British province of Bengkulu in Sumatra, in the middle of the nineteenth century.

Philippa Glanville, in *Silver in Tudor and Early Stuart England* (1990), talks of the seventeenth-century growth in fashionable accoutrements that accompanied the use of spice: spice boxes, graters, punch ladles, spice plates specially for serving spiced bread and candied fruit, and the pomander. This latter held spices and perfumes and was carried around to ward off noxious smells. It was also thought to offer protection against infections such as the plague. Silver nutmeg graters were a seventeenth-century fashion in England (Plate 7), the cookery writer Elizabeth David noting that 'no fastidious traveller need ever have been without a nutmeg to grate upon his food, his punch, his mulled wine, his hot ale or comforting posset' (David, 1970). A nutmeg grater made from a brown-spotted tropical stag cowrie (*Cypraea cervus*) (Plate 8), was probably an export from the Indies as was the nutmeg it grated. A silver grating-plate covers the mouth of the

7. A seventeenth-century English nutmeg grater. (Philippa Glanville, Victoria and Albert Museum, London.)

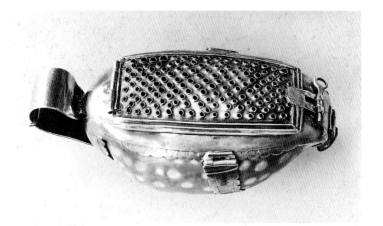

8. An eighteenth-century English nutmeg grater made from a cowrie shell. (Victoria and Albert Museum, London.)

shell. The broad loop at one end was possibly designed for attaching the grater to a belt or chain.

When he visited the Bandas in 1857, Wallace wrote of 'the small size and isolated position of these little islands … the chief nutmeg-garden in the world … almost the whole surface is planted with nutmegs, grown under the shade of lofty Kanary trees (Kanarium commune) [the Indonesian almond]. The light volcanic soil, the shade, and the excessive moisture of these islands, where it rains more or less every month of the year, seem exactly to suit the nutmeg-tree … few cultivated plants are more beautiful.' He noted, too, the large lilac-blue nutmeg pigeon, with its distinctive booming call, which feeds off the nutmeg. Digesting the mace, which gives the bird's flesh a delicate flavour and colour, the bird then voids the nut.

Nutmegs are harvested in June and December. Pickers use a lozenge-shaped basket similar to that used for hundreds of years by the Bandanese (Plate 9). This basket, of rattan, has tongs for hooking the fruit off the branch. Fruits are also commonly collected as they fall to the ground when ripe. Yields vary, though they can reach as many as 1,500 fruits from a single tree in a good year.

9. Basket used for picking nutmeg in Banda. (Neil Brierley.)

The nut, with its coating of mace, is removed by hand from the outer covering of fruit, and then the mace is carefully peeled off the nut. Dried in the sun, it loses its crimson hue and turns a browny-red, a process which in Indonesia's hot sun takes only a few hours. After the removal of the mace, the nutmeg is also sun-dried to prevent fermentation, and then shelled before being graded according to size. Nutmeg is then exported either whole or ground.

Nutmeg oil, which is extracted from the seed, leaves, and bark, adds taste to carbonated soft drinks and is widely utilized in the flavouring of canned and processed foods. Powerfully aromatic, even minute amounts can mask unpleasant odours. This explains its ancient use as a fumigant—it perfumed the streets of Rome before the coronation of Emperor Henry VI (Rosengarten, 1969)—and is employed in modern times in aerosols and household sprays. In small quantities it contributes its warm spicy notes to expensive perfumes, as does the highly priced oil of mace. Nutmeg oil is used mainly as a drug in the East, where it is valued for its stimulative, carminative, and aphrodisiac properties (Burkill, 1966).

Indonesia remains the world's main producer of nutmeg and mace. As with cloves, their production is now more evenly distributed throughout the archipelago: Sulawesi, West Sumatra, and West Java are producers in addition to the Banda Islands.

Pepper Piper Nigrum Merica

Pepper is one of the world's oldest spices. It was introduced into the Indonesian islands of Java and Sumatra around AD 600 from southern India, where it had spiced curries before the introduction of chillies in the sixteenth century. Its Indonesian name is *merica*, direct from the Sanskrit, as are the names of so many traded spices.

Pepper was known to the Greeks and Romans as a medicinal as well as culinary spice. Early references to it were made by Aristotle and Pliny, and Hippocrates in the fourth century BC recommended it mixed with honey and vinegar as a medication (J. I. Miller, 1969). The pepperers and spicers of twelfth-century Europe formed guilds to control the trade, which were later to become the apothecaries where spices were dispensed as medicines.

Pepper was so valuable a commodity that it was used as currency, hence the term 'a peppercorn rent'. It paid taxes, formed part of wedding dowries, and seasoned the monotonous Western diet of the Middle Ages. Pepper disguised the taste of heavily salted meat which had to last all winter, and masked the taste of putrid food: 'Together with other spices it helped to overcome the odours of bad food and unwashed humanity' (Purseglove et al., 1981).

In the seventh century, there began a highly profitable pepper trade between the Indonesian island of Sumatra and China, which grew in response to the enormous Chinese demand for the spice. By the sixteenth century, this trade had shifted to the port of Banten in West Java, which controlled the pepper grown in the Javanese hinterland as well as much of Sumatra's production

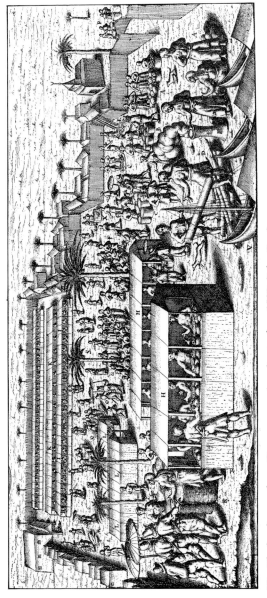

10. The bustling market at Banten, engraving, Amsterdam, 1598. (John Guy.)

of the spice. Banten was a centre of Asian trading, and most merchants had settlements in the town. The largest belonged to the Chinese, who exported 1,500 tons of pepper a year to China. The daily market in Banten, held outside the city gates, bustled with the activity of merchants (Plate 10): 'Gujerati and Bengali with painted articles and trinkets, Persians and Arabs with jewels, rows of Chinese shops with all their expensive goods: damask,

11. A pepper vine, *Piper nigrum*, from William Marsden, *The History of Sumatra*, London, 1738.

velvet, satin, silk, porcelain, lacquered work, medicinal products ..., and the like. Here, then, was the exchange, the meeting place of merchant gentlemen and ships' captains' (Van Leur, 1955).

Pepper was the most lucrative spice to be traded by the Portuguese and the Dutch during their years in the Indies. At the beginning of the seventeenth century, annual European consumption was around 3,000 tons. America entered the pepper trade in the middle of the eighteenth century, when her clipper ships began regular sailing between Salem, Massachusetts, and Batavia, the Dutch capital city on Java. The profits from this trade are said to have made America's first millionaires.

Piper nigrum, from which we get both black and white pepper, is a climbing vine with dark glossy leaves whose fruits—pepper—grow in elongated clusters (Plate 11). These fruits are green before they ripen to red, but turn black when dried. One cluster can produce as many as fifty fruits. To obtain the black pepper of commerce, berries are picked before they are ripe and dried in the sun. To obtain white pepper, the larger berries are left on the vine till ripe, then picked and the outer skin peeled off, revealing the seed which is then dried in the same way as black pepper.

Today, pepper remains the most commonly used spice in the Western world, and Indonesia is still a main supplier: over 50 000 tonnes were exported in 1991. As well as its wide domestic use as a condiment, pepper seasons commercially prepared snacks and instant foods. In tiny amounts it is an important ingredient in several well-known women's perfumes and, as a more masculine spice, is used also in men's cosmetic products.

Cinnamon and Cassia Cinnamomum burmannii
C. cassia Kayu manis

Both cinnamon and cassia trees produce a perfumed bark, classically defined as an aromatic, that is one of the oldest spices known to man. Called in Indonesia *kayu manis*, or 'sweet wood', it is in fact the bark and not the wood that is the spice (Colour Plate 2).

29

Cinnamon and cassia are grown in West Sumatra and Java, from where they have been traded for 2,000 years. They were, in fact, among the first spices searched out by fifteenth-century European explorers, and appear to have been traded interchangeably, though the bark of cassia is coarser and cinnamon is generally considered superior. Burkill (1966) mentions that throughout history cassia has been regarded as a second-class cinnamon bark.

The term cinnamon generically covers both woods, which belong to the laurel (*Lauraceae*) family of over 2,000 species. These include the clove, the sweet bay (the laurel of the Romans) whose leaves are used in Western cooking, and camphor. The stately Barus camphor (*Dryobalanops aromatica*), native to Barus on Sumatra's inhospitable west coast, produces astringently aromatic crystals that were a medicinal spice the Arabs called Luban al-Jawi, or Incense of Java. It was the Arabs who carried this spice to the Mediterranean, and there was vigorous commerce in camphor with India and north to China from the beginning of the first millennium. Famed throughout Asia as a drug, it was successfully substituted by Sumatran traders for Persian frankincense in the China trade, in the same way that the resin benzoin was substituted for myrrh.

Cinnamon was traded at the same time as camphor. Several references attest to its early use: the Lord instructed Moses, in the Old Testament of the Bible, to use cinnamon and cassia to anoint the tabernacle. Nero is said to have burned it during his wife's funeral in 66 AD, and Chinese herbals of this period describe its medicinal properties. So great was the demand for cinnamon in Egypt—it was an important element in embalming—that Queen Hatshepsut mounted her own expedition to Somalia to collect the spice in 1485 BC.

Cinnamon trees grow as tall as 17 metres in the wild, but under cultivation the constant removal of shoots by cropping results in a dense bush around 2 metres tall. Harvesting is done during the rains when the bark lifts easily, and the red flush of the young leaves (a laurel trait, and found with the clove, too) is turning to green. On a limited scale, harvesting can be done throughout the year.

To gather the spice, a small knife is used to ease the inner bark off the wood in strips about 1 metre long. These are then dried, curling into quills and turning the familiar red-brown. The cinnamon is then broken into small lengths to produce the neat rolls that are on sale in the West, where its modern use is as a popular kitchen flavouring. Thousands of litres of bark oil, or oil distilled from the leaves, which are also strongly aromatic, as are the twigs, are exported from Indonesia. This oil provides the warm, spicy notes of some fine perfumes, adds flavour to processed foods, and contributes its distinctive flavour to several carbonated drinks. There is, as well, a small export trade in dried cassia buds, which are also used as a spice.

Ginger *Zingiber officinale Jahe*

Turmeric, ginger, galangal, and cardamom all belong to the family *Zingiberaceae*. Although lacking the exotic connotations of the better known clove, nutmeg, and cinnamon, members of the humble ginger family have been traded and used in the East, and carried to the West, for over 2,000 years.

In the East, where it is known for its stimulating and carminative properties, ginger has always been widely employed both as a culinary spice and as a medicine. In the fifth century AD, it was transported, growing in pots, by Chinese and Indonesian sailors, who ate it on board to ward off scurvy. In the eighteenth century, the German naturalist Georg Everhard Rumphius, who was to document the flora of the Moluccas during his years there in the service of the Dutch East India Company, remarked that there was hardly a prescription in Moluccan medicine that did not employ it.

One of the first Oriental spices to be used in Europe, ginger may have been carried along with cinnamon to the East African coast in the first century AD. It was described at this time by Dioscorides, the author of an early medical manual, as being 'whitish in colour and peppery in taste, and of a fragrant smell'. Known in England before the Norman conquest, it appears to have been as readily available as pepper, but less expensive;

1 pound was the price of a sheep. Fifteenth-century English cookbooks seemed to require it in nearly every recipe. Henry VIII recommended its use against the plague, while in the court of Queen Elizabeth I, gingerbread was a popular confection (Purseglove et al., 1981).

The ginger of commerce is the aromatic rhizome of a slender perennial with grass-like leaves, thought to originate from East Java (Colour Plate 3). Although it is the fresh rhizome that is mostly consumed in the East, for export, ginger is generally dried, as are most spices, for preservation and easy transportation. Ground ginger provides the commonly used spice in the West, where it is added to cakes and cookies, sauces and pickles, ginger ales and beer, and is a main ingredient in commercially prepared curry powders. Ginger is also exported preserved and crystallized, as is ginger oil made from the dried rhizome. This oil is employed in perfumery, pharmaceuticals, and as a flavouring. Indonesia, in addition to growing large quantities of ginger for domestic consumption, exported over 46 000 tonnes in 1991.

Turmeric Curcuma domestica Kunyit

Turmeric is a bright orange rhizome belonging to the ginger family that has been cultivated and traded throughout the East for thousands of years. The sanctity of its colour is traced by Burkill (1966) to the ancient cult of sun-worshipping, in which the golden-red, sun-coloured saffron was an object of veneration. Turmeric, easily grown and widely available, was a cheap alternative. Its name in Indonesia and throughout the Malay Peninsula—*kunyit*—is the word for 'yellow'. The botanical term *Curcuma* is derived from *kurkum*, the name given by the Arabs who traded the rhizome.

While turmeric is not a drug as are other spices, it is used in traditional medicine in the East. It is also employed as a dyeing agent, providing shades of orange and yellowy-brown to silks, cotton, and food, and as a cosmetic colourant in ceremonies. With its slightly bitter, spicy taste, it is widely consumed as a condiment in the East, where it is the freshly harvested rhizomes

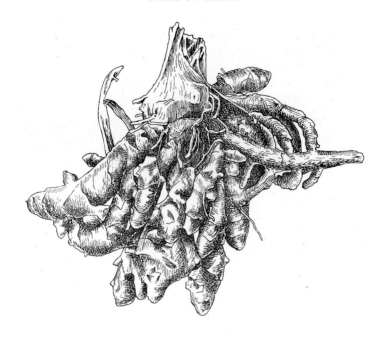

12. Turmeric, *Curcuma domestica*, from *Bulletin du Jardin Botanique de Buiten-zorg*, 1918. (Royal Botanic Garden, Edinburgh.)

that are used (Plate 12). In the West, which imports it mainly dried or ground, it is an important ingredient in the colouring and flavouring of commercial curry powders and packaged snack foods; it is often utilized more for its colour than its flavour.

Galangal (Greater) *Languas galanga* Laos

The aromatic rhizome of this tall member of the *Zingiberaceae* family—the plant often grows higher than a man—is a spice of Eastern commerce that was described by the Chinese as 'mild ginger'. Their annals of the northern Wei Dynasty (AD 385–557) referred to galangal as an article of trade from the Po'ssi, or South (J. I. Miller, 1969). Burkill (1966) writes that it was used as a medicine in Mesopotamia in the sixth century, and was well known in Europe by the Middle Ages. Widely used in food preparation, galangal was known in the West as galingale.

The spice is native to Indonesia, where it is called *laos*, and from where it has been traded for thousands of years. It is as extensively cultivated as ginger and turmeric throughout South-East Asia, the pinky-yellow rhizome being commonly used in daily cooking.

Seeds of galangal (Colour Plate 4) are used locally as cardamom, though Indonesia does grow species of the true spice, *Elettaria cardamomum*, and the related *Ammomum* variety, which are traded locally (Colour Plate 5). In the East, cardamom is chewed to sweeten the breath, and is sometimes included as a flavouring with the areca nut chew. In the Arab world, the spice is served as a flavouring for coffee, while in the West, where cardamom also has culinary applications, the oil is an important fragrance in some fine perfumes.

5
The Portuguese in the Indies

ON the eve of European arrival in the Indies in the sixteenth century, the focus of Asian trade had shifted from the Indonesian kingdoms of Srivijaya and Majapahit to Malacca on the Malay Peninsula (Plate 13). Malacca controlled the Strait and was the major entrepôt in the area. Her cosmopolitan port is said by early travellers to have harboured more ships than any other place in the world (Plate 14). Here, merchants refurbished their ships and traded their wares while they waited for the next monsoon to take them on to the Indies and China, or back to home ports. Tomé Pires wrote of the port, in 1515, that 'No trading port as large as Malacca is known, nor any where they deal in such fine and highly prized merchandise. Goods from all over the east are found here; goods from all over the west are sold here. It is at the end of the monsoons, where you find what you want, and sometimes more than you are looking for.' He told of the Arab traders who brought to Malacca the riches of Venice: 'coral ... silver, glass and other beads, and golden glassware'. These traders collected, in return, the spices of the Moluccas, carrying them back to the markets of the Mediterranean.

It was via trade from Malacca, whose rulers had converted to the Muslim faith in the fourteenth century, that Islam reached Java and spread east to Ternate and Tidore before 1500. It was these same clove-producing islands which were to be Catholic Portugal's first contact with the East Indies. Portugal's sixteenth-century search for spice riches was to be combined with a crusade against the Muslim heathen. It was thus a quest for both God and Mammon.

The Portuguese were a race of tough fighters and seamen. They had crusaded for centuries against the Muslim Moors of Morocco, finally driving them from their Portuguese homeland on the Iberian Peninsula in 1415. As seafarers, life on the harsh

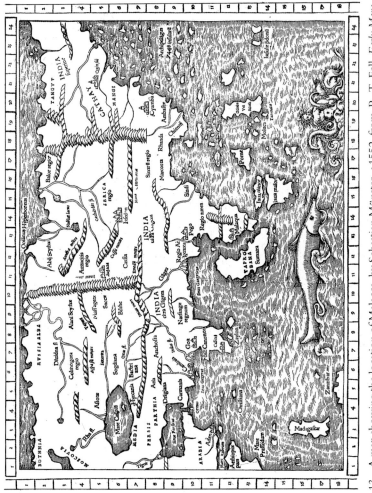

13. A map showing the location of Malacca, by Sebastian Münster, c.1552, from R. T. Fell, *Early Maps of South-East Asia*, Singapore, Oxford University Press, 1988.

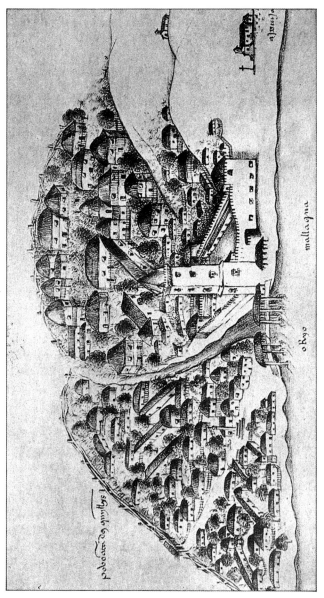

14. Portuguese Malacca, c.1540, from Gaspar Correia, *Lendas da India*, from R. T. Fell, *Early Maps of South-East Asia*, Singapore, Oxford University Press, 1988.

37

north Atlantic sea coast, from where they traded and fished, had equipped them with an innate knowledge of the sea, winds, and tides. They had developed ship-building and nautical skills which enabled them to plot maps, read nautical charts, and use a basic astrolabe and marine compass. These skills were guarded jealously from the rest of Europe, thus enabling Portugal, in the sixteenth century, to dominate more of the world and its trade than any other country (Boxer, 1969).

Prince Henry the Navigator, an enthusiastic crusador against the Moors, gave continuous encouragement to his seafaring Portuguese nation in their voyages of exploration. He recruited knowledgeable mariners and geographers and opened a school of maritime studies, which led to the growth of Portugal's surprisingly strong and well-armed navy. At the height of her power in the sixteenth century, Portugal had 300 ships, which were to prove far superior to the junks and dhows of the Indian Ocean. Utilizing this fleet, the Portuguese were the first European nation to explore the sea route to the East, where they developed trade links and established an empire.

In their search for a southerly passage to the East, the Portuguese initially pressed cautiously down the west coast of Africa. They were encouraged in this route by the prospect of finding slaves and gold as well as a way East. Information on a trade in the precious metal had been gleaned during Portugal's crusading campaigns in Morocco, from where West African gold was traded to the merchants of the Mediterranean. While they never found the source of gold, the Portuguese did manage to divert a proportion of the trans-Sahara trade in the metal to their own ships and coastal fortifications (Boxer, 1969).

Progress down the inhospitable west coast was slow. It was not until 1488 that Bartolomeu Dias rounded the Cape of Good Hope, sailing as far as the Persian Gulf and returning to Portugal with the news that the sea route to the Indies was open. Vasco da Gama followed him round the Cape in 1497, reaching the Indian Ocean and Calicut (the most important Indian port of the spice trade) on India's west coast. Da Gama arrived back in Lisbon with the first Eastern cargo brought to Europe by sea, and with

3cm

1. Nutmeg, *Myristica fragrans*, from W. Veevers-Carter, *Riches of the Rain Forest*, Singapore, Oxford University Press, 1984.

The Java Cinnamon.

2. Java Cinnamon, showing a piece of the cut bark, 1760. (Antiques of the Orient, Singapore.)

Zingiberaceae.

Zingiber officinale. Rosc.

172

3. Ginger, *Zingiber officinale*, from Eugen Kohler, *Medizinal-Pflanzen*, Gera-Untermhaus, 1887. (Royal Botanic Garden, Edinburgh.)

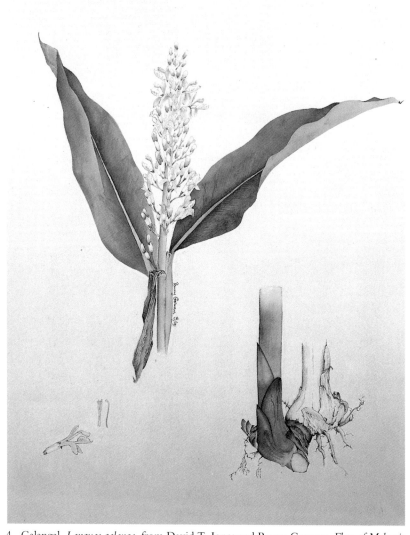

4. Galangal, *Languas galanga*, from David T. Jones and Penny German, *Flora of Malaysia Illustrated*, Kuala Lumpur, Oxford University Press, 1993.

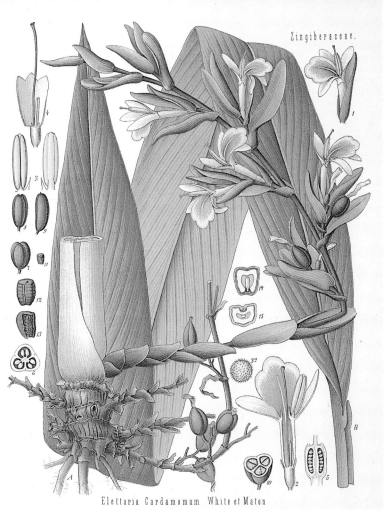

Zingiberaceae.

Elettaria Cardamomum White et Maton

5. Cardamom, *Elettaria cardamomum*, from Eugen Kohler, *Medizinal-Pflanzen*, Gera-Untermhaus, 1887. (Royal Botanic Garden, Edinburgh.)

6. Foreign traders being welcomed at Ternate, engraving by De Constantin, from *Rescueil des Voyages qui out Servi a L'Etablissement et aux Progres de la Compagnie des Indes Orientales*, Amsterdam, 1702. (Antiques of the Orient, Singapore.)

7. Banda, showing the volcano Gunung Api on the left and Banda Neira and the Dutch Fort Nassau on the right, lithograph by J. C. Rappard, from M. T. H. Perelaer, *Het Kamerlid van Berkenstein in Nederlandsch-Indie*, Leiden, 1888–9. (Collection of the author.)

8. View of the waterfront at Batavia, from W. Schouten, *Oost-Indische Voyagie*, Amsterdam, 1676. (Antiques of the Orient, Singapore.)

De stadt.
BATAVIA

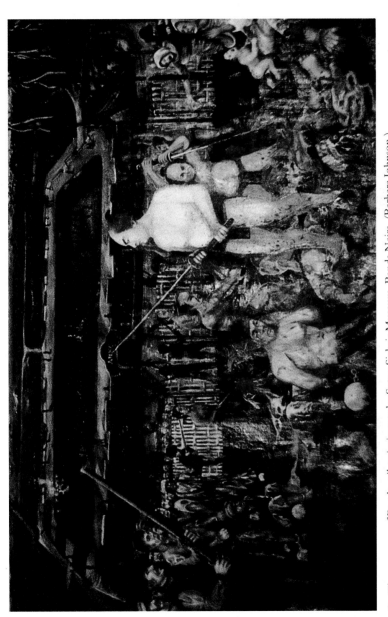

9. The massacre of Banda, oil-painting at the Sutan Sjahrir Museum, Banda Neira. (Barbara Johnson.)

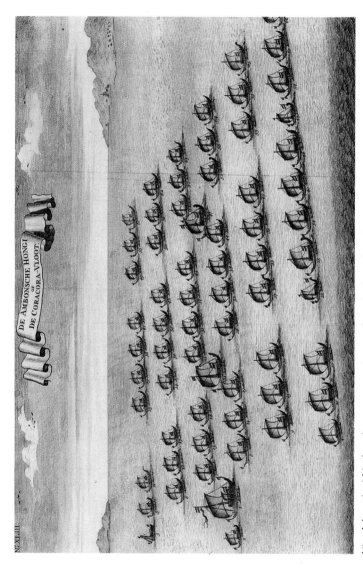

10. A hongi raid in *kora-kora* canoes setting out from Ambon, from François Valentijn, *Oud en Nieuw Oost-Indiën*, Dordrecht, 1724–6. (Antiques of the Orient, Singapore.)

11. The country home of Sir Thomas Stamford Raffles at Pemantang Balam, outside Bengkulu, c.1823, from Mildred Archer and John Bastin, *Raffles Drawings in the India Office Library, London*, Kuala Lumpur, Oxford University Press, 1978.

12. Rolling *kretek* in a large factory. (PT Gudang Garam, Jakarta.)

13. Model of a sailing vessel being made from strung cloves. (Mochtar Lubis.)

14. A cone of *nasi kuning*, or yellow rice, being sliced. (PT Sucofindo, Jakarta.)

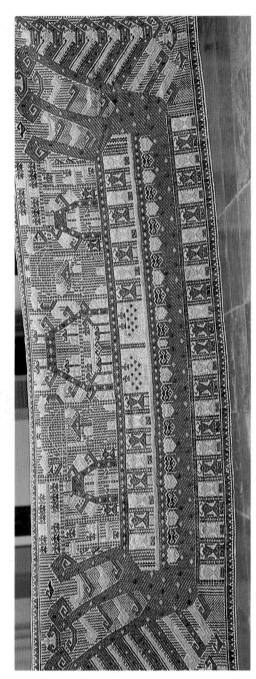

15. A nineteenth-century ship's cloth from the Lampung region of south-west Sumatra, showing the use of turmeric as a dye, from the collection of the National Museum, Jakarta. (Hilde Ardie.)

16. The *bunga cengkeh*, or clove motif, is here used to form chains on the broad end border of a sarong and to embellish the spaces surrounding the triangular 'cockerel's tail feathers' motif, from Grace Inpam Selvanayagam, *Songket: Malaysia's Woven Treasure*, Singapore, Oxford University Press, 1990.

17. *Areca catechu* palm, from Eugen Kohler, *Medizinal-Pflanzen*, Gera-Untermhaus, 1887. (Royal Botanic Garden, Edinburgh.)

information of a highly developed web of local trade and shipping which he had found established in the Asian seas (Map 2).

Portugal's first settlement in the East was established by Francisco de Almeida at Cochin on India's Malabar coast, where de Almeida grandly styled himself Viceroy of India. From this base, his successor, Alfonso de Albuquerque, captured Goa in 1510, which was to remain in Portuguese hands for the next 450 years. With Goa and the Indian Ocean under Portuguese domination, de Albuquerque imagined he had control of the spice trade. But this remained in the hands of the thriving emporium of Malacca on the Strait, which was still used by Muslim merchants. Until Portugal gained command of this strategically placed port, she was not going to succeed in a trade monopoly.

After a lengthy blockade, Malacca finally fell to the Portuguese in 1511, but the spice monopoly hoped for was never completely achieved. Portugal's occupation was constantly threatened by the strongly Muslim population of Aceh in North Sumatra, who continued to defy the Portuguese by shipping thousands of tons of pepper to Mecca in the Red Sea in the sixteenth century.

From newly conquered Malacca, the Portuguese lost no time in dispatching an expedition to find the eastern source of spice. This was led by Antonio d'Abreu, still recovering from a vicious face wound received in the battle for Malacca. Even though d'Abreu lost two of his three ships in a storm at sea, and was unable to find the northerly clove islands, he did manage to reach Banda where he took on nutmeg and mace. He had instructions to respect local customs and attempt friendly relations, and on his return he reported on the simple lives of the people he had seen, the women who worked in the nutmeg groves, and the scented trees whose fruits they traded for rice and cloth.

Further expeditions were more successful, culminating in 1513 when the northern Moluccan clove islands were reached. Portuguese influence was to be strongest here, in tiny, volcanic, clove-producing Ternate and Tidore. The two sultanates, rivals for control of the region and the clove trade, vied against each other for Portuguese support. The Portuguese chose the more powerful Sultanate of Ternate, and permission was granted by its

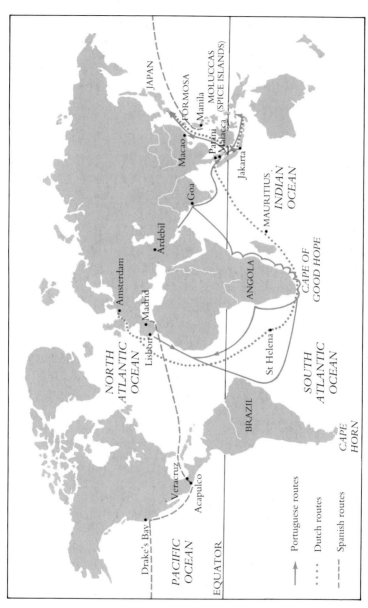

Map 2. European sea trade routes in the sixteenth and seventeenth centuries, after Maura Rinaldi, *Kraak Porcelain: The History and Classification of Dishes*, London, Bamboo Publishing, 1989.

Sultan to establish a trading post, known as a factory. The Portuguese were also allowed a large clove cargo (Colour Plate 6).

At this stage, a Spanish-sponsored expedition under the Portuguese Ferdinand Magellan reached the Moluccas. Magellan had sailed from Spain in 1519, down the South American coast, through the Strait that is named after him, and across the Pacific to reach the southern Philippines in 1521. From there, the convoy sailed to Tidore, though Magellan was to die before he reached the island, killed in a fight with natives on Mactan Island in the Philippines. But his expedition had achieved a westerly route to the Spice Islands, and the Tidorese were only too glad to welcome a Spanish presence on their island to support their struggle against Portuguese-occupied Ternate.

On his voyage, Magellan had been accompanied by an Italian, Pigafetta, who wrote of the expedition's brave search for the Moluccas on a journey that took twenty-seven months and two days and who studied and recorded in his diary the rare spices they found. Only one vessel returned to Spain in 1522, via the Cape of Good Hope, and just eighteen of the original crew of 270 who had sailed from Seville in 1519.

For several years after Magellan's pioneering voyage and the establishment of a Spanish base on Tidore, Spain was to challenge Portugal's attempted monopoly of the Indonesian spice trade. Portugal claimed Spain was violating the Treaty of Tordesillas. By the terms of this Treaty, Pope Alexander II had attempted, as early as 1494, to resolve territorial rivalry between the two Catholic neighbours who shared the Iberian Peninsula by dividing the unconquered world between them. All lands to the west of an imaginary line drawn 370 leagues west of the Cape Verde Islands went to Spain, who by its terms was granted America (discovered by Christopher Columbus in 1492 in his search for the Spice Islands). Those lands to the east of the line went to Portugal. But it was not until Spain colonized the Philippines in 1570 that she relinquished her base in the Moluccas.

Though Portugal achieved a foothold in the clove market, her efforts at monopolizing the nutmeg trade of Banda were never successful. The cost of equipping large carracks—the capacious

Portuguese merchant ships—to reach both the southerly Bandas for nutmeg and mace and the northern Moluccas for cloves, was a heavy drain on finances, as was maintaining forts and factories in both areas. A start was made in 1529 on building a fort on the Banda island of Neira, but it was never completed. The Bandanese, also, were not keen to do business with the Portuguese. They retained their close trading connections with the Muslim sultans of Java, built up over years of interisland commerce. Specific Portuguese trading in nutmeg therefore stopped in 1561. Supplies of the spice were obtained instead from regional traders, who carried it to the Portuguese base in Ternate.

Portugal's initial onslaught on eastern riches has been chronicled as piratical (Boxer, 1969). With an old-fashioned and feudalistic administration, the Portuguese were 'better warriors than philosophers, and more eager to conquer nations than to explore their manners or antiquities' (Marsden, 1738). Ternate's Governor Menesez was arrested in 1530 by his successor for corruption and cruelty to the islanders and sent to Lisbon in chains—not an uncommon occurrence among Portuguese administrators. In contrast was the respect and regard in which Governor Antonio Galvão (1536–40) was held. His *História das Molucas*, an important record of life at that time, evocatively describes the trade, islands, people, and spices (Van Leur, 1955).

Though the Portuguese undoubtedly disrupted the ancient web of peaceful Asian commerce, they did not manage to totally destroy it. They were unable to enforce a monopoly on local trade, and managed only a share in the profits. In the second half of the sixteenth century, however, these were substantial, with huge sums being made on early voyages. European demand expanded, spice prices trebled, and production increased to meet the demand. In fact, more cloves were grown than could be exported and these were left unpicked on the trees.

Spices in Europe were literally worth their weight in gold, which was the only acceptable currency of the spice trade. Portugal obtained part of her gold supplies from what was known as the 'gold island' of Sumatra. The gold was probably minted in Cochin, producing coins known as the sao tome, which were traded in the markets of the East alongside the Venetian ducat and the Indian mohur.

Portuguese carracks annually plied the route home round the Cape of Good Hope to the major European distribution centre of Lisbon. Between 1571 and 1610, they carried average annual spice cargoes of over 1,000 tons of cinnamon, cloves, nutmeg, ginger, and most importantly, pepper from Sumatra and Java. Pepper formed well over half the spice cargo and was to bring Portugal the most lucrative returns. Indeed, it was the only spice which made a sizeable profit for the crown, who took one-third of all profits. These were to fall sharply in the seventeenth century with competition from the newly arrived Dutch and British traders.

The growth of Islam was a constant threat to Portugal's expansion in the archipelago, and an attempt was made to stem its spread by converting the population to Catholicism. During the sixteenth century, the Jesuits, many of whom were in the East to make a profit for themselves as well as to spread the gospel, made concerted missionary efforts. These were concentrated mainly on the island of Ambon, north of Banda. The majority of Portuguese converts were to be made here, even though Ambon's Hitu province had a Javanese colony which offered strong Muslim resistance. It was thought by the Portuguese that Ambon would be a useful southerly base after Ternate in the north, where they dared not threaten their lucrative clove trade with the Sultan by conducting Christian missionary activity.

By the middle of the sixteenth century, Portugal had possessions in Brazil, on the west coast of Africa, at Morocco, and in the East, where her bases on the Asian sea routes stretched from Ormuz in the Persian Gulf to the Moluccas and Macao (in 1557). To control and administer this far-flung Empire, thousands of young men sailed every year from Portugal. Few were to return, which was an enormous drain on the country's manpower. Many would die on the long voyage. Others would be killed in battles at sea, some would desert, while large numbers would be decimated by disease—malaria, dysentery, and fevers.

By the end of the sixteenth century, Portugal's sea power was

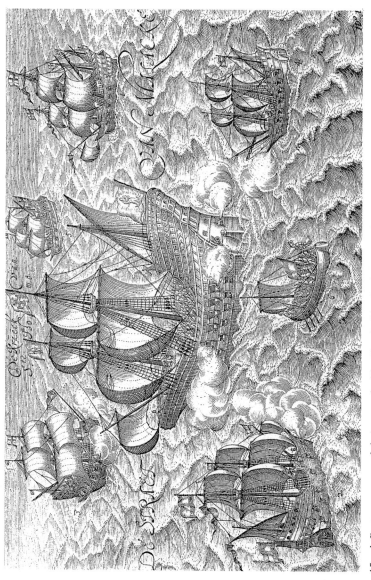

15. A Portuguese carrack being attacked by Dutch ships in the Strait of Malacca, from Spilbergen, *Historiael Journael*. (Library of Congress.)

proving inadequate and her navy too small to patrol her Empire. With not enough ships or men of her own to man her overseas service, Gujerati Muslims—accomplished sailors—often crewed Portuguese vessels on their Asian routes. However, the expense of fitting out these annual carracks, maintaining often dishonestly administered forts in the Moluccas, as well as trying to compete with successful local smuggling, resulted in heavy losses to the Crown in Lisbon. Contributing to this drain on Portuguese reserves was the private trading in spices, supposedly controlled, that was condoned by the Crown to supplement the inadequate salaries of officers.

The Portuguese retained a base on Ternate until 1570, when Sultan Hairun was treacherously killed by the Portuguese after an argument over his share of the clove profits. Led by Hairun's son Baabullah, the whole of Ternate rose in revolt, besieging the Portuguese fort on the island for the next five years. The Portuguese were forced to look for an alternative base, which was offered by Ternate's rival Tidore, and a factory was established there in 1578.

The British now made an appearance in the form of Sir Francis Drake. On his arrival in Ternate, he was offered a lading of spices and a treaty by the new Sultan of Ternate, Baabullah, who was anxious for British assistance against the Portuguese who had murdered his father. On his return to England, great interest was shown in Drake's voyage, and this provided an impetus to the British to investigate for themselves the spice source of the Indies.

When the British appeared on the scene, it was to take them, and their Dutch rivals, a surprisingly long time—almost sixty years—to wrest control of the spice trade from a declining Portugal. Dutch expansion in the seventeenth century saw Portugal driven from the northern Moluccas and, against fierce resistance, from Malacca (Plate 15). Although the Portuguese managed to retain a trading base at Ujung Pandang in southern Sulawesi, they were eventually driven from there too, in 1667. At the end of the seventeenth century, Portugal's influence was at an end in the Indonesian archipelago.

6
The Dutch and the British in the Indies

HOLLAND was eventually to take control of the spice trade from Portugal. This she was able to do, first, because of her superior sea power; the stoutly constructed Dutch Indiamen outstripped the lumbering Portuguese carracks, which had not been modernized or updated since the sixteenth century, nor their fleet augmented. Secondly, Holland had economic resources superior to Portugal. These had been accumulated by the Dutch as the tradesmen of Europe, carrying and trading goods around northern Europe and Scandinavia. Additional manpower for Holland's overseas service (the number of foreigners serving in the Dutch colonies was always considerable) was later provided by these European neighbours.

In the seventeenth century, the three European powers in Asia—Portugal, Spain, and Holland—were all ruled by Spain. Holland was struggling for independence from the Spanish Crown (not to be granted until 1648) when Spain annexed neighbouring Portugal in 1580, and in 1595 shut down the port of Lisbon, and thus Europe's spice centre, to her rebellious Dutch subjects (McKay, 1976). This was motivation for Holland to seek out for herself the source of spice.

Original Dutch enterprise in the East was in the form of individual trading companies, unlike the Portuguese enterprise which had been sponsored by the Crown. The first voyage, in 1595, under the command of Cornelius de Houtman, was to signal the start of over 300 years of Dutch involvement in the Indies. De Houtman sailed round the Cape, across the Indian Ocean, and through the Sunda Strait (see front endpaper). Losing one ship and 145 men out of his original crew of 249, he was thrown into prison in the port of Banten in Java for what the historian Hall (1964) called his 'outrageous behaviour'. De Houtman was ransomed a month later, and allowed to travel on to Jayakarta (later to become the Dutch capital, Batavia, and modern Jakarta). From there he sailed east to Bali, where he was forced by his officers to return home to Holland without attempting to reach the Moluccas. But even the small cargo of spices he had managed to obtain on the trip was received with jubilation in Holland.

At the same time as the Dutch, the English had been searching for a route to the East. As early as the fifteenth century, during the reign of King Henry VII, England had been unsuccessfully exploring a northern passage round America or Russia. By the end of the sixteenth century, the English realized that the only practical route East was round the Cape of Good Hope. They were held back in their efforts at exploration, however, by both a lack of funds and a supply of suitable trade goods; England knew that the tropical East would not provide a market for her heavy woollen cloth. Encouraged by the success of Francis Drake's expedition with its spice cargo from Ternate, and the plundered riches of a Portuguese carrack, then by the English defeat of the Spanish Armada in 1588, London merchants petitioned Queen Elizabeth I to sanction an expedition.

Shortly after, in 1591, George Raymond and James Lancaster sailed from Plymouth. Lancaster, like the Dutchman de Houtman, had spent time in Lisbon studying Portuguese navigational techniques. Nevertheless, this expedition was to suffer appalling mortalities and the loss of Raymond's ship off the Cape. Lancaster was able to sail on to north-west Sumatra, managing some profitable plundering of Portuguese shipping in the Strait of Malacca on the way, and establishing for the British a trading post in Banten.

In 1598, Jan Huygen van Linschoten, a Dutchman who had also sailed and studied the ways of the sea from the Portuguese in Lisbon and Goa, published his *Itinerario*. This provided invaluable information on the trade and navigation of the Indian Ocean, and of the southerly route to the Indies via the Cape of Good Hope. It prompted the English to form a company to trade in the East, to consolidate their post at Banten, and to challenge the Dutch attempt at a spice monopoly in the Indies. It also provided the next Dutch expedition with much needed maps and navigational charts for their venture.

In the year the *Itinerario* was published, five expeditions, numbering twenty-two ships, left Holland for the East. Throughout the area—in the Indies, Thailand, the Philippines, China, and Japan—these were usually warmly received and allowed to trade, and Dutch help was also sought against the Portuguese. An exception to these peaceful relations was the murder of Cornelius de Houtman in 1599 in Aceh, North Sumatra (it was he who had led the first Dutch attempt to find the Indies). His death was instigated by the Portuguese, anxious to be rid of Dutch trespassers, and was ordered by the Acehnese sultan who needed Portuguese assistance in his feuds with a neighbouring Malay state. Cornelius's brother Frederik, a noted linguist and the navigator of de Houtman's fleet, was imprisoned in Aceh. (During his incarceration he compiled the first Malay–Dutch Dictionary (Hall, 1964).)

The largest expedition to leave Holland for the East Indies, under the command of Jacob van Neck, Wybrand van Warwijck, and Jacob van Heemskerk, discovered Mauritius on the outward voyage and named it for Prince Maurice of Nassau. Van Neck made record time in reaching Banten on Java, and filled four ships with pepper before sailing home. The remaining ships in the convoy went east to the Moluccas, arriving at Ambon, the clove-producing island north of the Bandas, which they were able to take with little resistance from the Portuguese. They established a base and left a caretaker behind. This led to the signing of Holland's first substantial treaty with Ambon's ruler in 1600, giving the Dutch the right to all cloves produced on the island. It was the first of many similar treaties aimed at securing a trading monopoly. By 1605, the Dutch were well entrenched with forts and garrisons on the island that was to be their headquarters for the Moluccas, islands which were to feel the full brunt of Dutch imperialism for 350 years.

The day of the Dutch arrival in Banda in 1599 was 'engraved upon the Bandanese memory' says Willard Hanna in his book, *Indonesian Banda* (1991). With gifts for the islands' rulers, and by careful negotiating, van Heemskerk was allowed a nutmeg cargo and was given permission for the establishment of two trading posts on Neira and Lontar (Colour Plate 7). When all the ships of van Neck's convoy finally returned to Holland, a total profit of 400 per cent was declared on the voyage. No other expedition in this period of indiscriminate trading, when there was hardly a port in South-East Asia not visited by the Dutch, was to make a profit of this size.

To put an end to individual profiteering, the United East India Company—the Dutch Vereenigde Oostindische Compagnie, known as the VOC—was founded in March 1602. From then on, all trading was to be done through 'The Company' which was run by seventeen directors known as the Heeren Zeventien, or XVII. With a twenty-year charter, it had trading rights from the Cape of Good Hope to the Bering Sea. It could write its own treaties, declare war or peace, build and maintain forts, establish colonies, execute justice, and coin its own money. Wielding enormous power, it was a state within a state. At its inception it had ten times as much money in its coffers as its English rival, the English East India Company, which was granted a royal charter in December 1600 to organize trading expeditions to the East, with a fifteen-year trading monopoly in the area between the Cape of Good Hope and the Magellan Strait.

Trailing the Dutch round the Indies, the British, under Henry Middleton, were chased out of Ambon in 1604. Middleton then sailed up to Ternate, which had been captured by the Dutch in 1607 against fierce Portuguese opposition, and managed to secure a clove cargo, but Dutch hostility forced him to repair to the British post in Banten. At this stage, Admiral Verhoeff forced local rulers to sign monopoly treaties in favour of the VOC, and when Middleton returned in 1610, claiming a right to trade as a nation friendly to the Dutch, he was threatened with force. Escaping to the tiny Banda island of Ai, he secured a good cargo of nutmeg, left two caretakers behind to establish a British presence, and persuaded neighbouring Run to hoist the British flag as well. Three years later, Captain Thomas Best managed to establish a factory at Aceh in North Sumatra, and one on the eastern coast of the island at Jambi. From their base in Banten, trading posts were planted by the British in Jakarta and at Japara on Java's north coast in 1617.

With so much profit to be made, there were to be continued British attempts at gaining a foothold in the Moluccas. Devon seaman John Jourdain was given the task of spearheading the British struggle against the Dutch in the Indies, with the aim of establishing a British base in the Moluccas. Jourdain negotiated with Jan Pieterszoon Coen (Plate 16), Holland's administrator of the large island of Seram, north of Ambon, but Coen, later to become Governor-General of the Indies, was rigid in his protection of the Dutch monopoly and Seram's ruler too frightened of him to trade with the British. Jourdain was successful, however, in founding a factory for the British at Ujung Pandang in Sulawesi, an important half-way house between Java and the Moluccas. Rice grown here was supplied to the Moluccas, where it fed the nobility—commoners ate the staple sago—in exchange for spices. It was from Ujung Pandang, too, that the British traded the cotton cloth and opium collected from India, and much valued by the Moluccans. But Ujung Pandang was eventually taken by the Dutch in 1667.

The first Dutch Governor-General of the Indies, Pieter Booth, was appointed in 1609. At the same time, a Council of the Indies was set up to assist in governing Holland's Eastern interests, but it was Coen, Booth's successor in 1618, who was to be the real architect of the Dutch empire in the East. An accountant by profession, he was a clever but ruthless politician. As early as 1614 (before becoming Governor-General), he had recommended vast territorial expansion in the East at the expense of the Portuguese and Spanish, and with the aim of annihilating the increasingly worrying British presence on the Eastern seas. Upon becoming Governor-General, he was instructed by the Heeren XVII to expel all foreigners from wherever the Dutch traded and to search their ships, if necessary, for spices. The Dutch realized that competition would push up the purchase price and cause a glut in what could only be a limited market, thus greatly reducing their profits.

IAN PIETERSZOON KOEN
Gouverneur Generaal van Nederlands Indiën.

16. Jan Pieterszoon Coen, the founder of Dutch Batavia, from François Valentijn, *Oud en Nieuw Oost-Indiën*, Dordrecht, 1724–6. (Antiques of the Orient, Singapore.)

Though the Dutch had built a warehouse and some houses in Fort Jayakarta as early as 1610, the city was not taken by Coen until 1619 from its ruler, the Sultan of Banten, in whose province it was located. Coen renamed the city Batavia, a name which would remain for the next 300 years (Colour Plate 8). Batavia was to be the capital of the Dutch empire in the East and, much later, the Jakarta of modern-day Indonesia.

Pressing home his victory in Jakarta, Coen sailed on to Banten to challenge the British presence there. He attacked John Jourdain while the British captain was attempting to round up supporting forces, and Jourdain was disastrously shot by the Dutch under a flag of truce. Soon after, Jourdain's compatriot, Nathaniel Courthorpe, was killed by the Dutch in Banda, four ships were captured by Dutch forces off the coast of Sumatra, and another in the Sunda Strait.

It was into this warlike scene that the unlikely news arrived of an Anglo-Dutch treaty, signed between the two countries in February 1619 in London. Spice profits were to be shared between England and Holland, with the English getting one-third and an equal share in the pepper profits of Banten. However, Governor-General Coen was to pay mere lip service to the terms agreed by his government, who had felt it was time to make peace with the English. The VOC was too powerful a body to regard any orders from the Dutch government. Showing immediate disregard for the treaty and any attempt to co-operate, the Dutch rounded up the staff of the English factory in Ambon, which had been allowed a limited trade in spices, forced them under torture to confess to charges of conspiracy against the Dutch, and executed them.

With the excuse that the Bandanese were not abiding by the terms of their trading treaty with the Dutch, Coen decided to complete a thorough conquest of the nutmeg-producing islands. He aimed to consolidate Dutch interests and enforce a monopoly in line with the expressed policy of the Heeren XVII in Holland: 'Banda and the Moluccas are the principal target at which we shoot.' Governor-General Coen arrived in Banda in March 1621 with a mixed army of Europeans, Javanese convicts, and Japanese mercenaries. Forcing village headmen on Lontar Island to admit, under torture, to trumped-up charges of conspiracy, he used their confessions of guilt to wipe out the Bandanese. Almost the entire population was killed, tortured, or sent into slavery on Java (Colour Plate 9). Some fled to the hills, to die later of exposure or starvation; others escaped to neighbouring islands where Coen dispatched ships to harass and burn villages. On Run, where the inhabitants tried to flee when they heard of the atrocities on neighbouring Lontar, all adult men were executed. For his decimation of Banda, Coen received only a mild public reproof from the Heeren XVII, despite the outcry his atrocities caused in Holland. Even in this year of decimation, 1621, VOC nutmeg sales to Europe were around 200 tons (Ellen, 1979).

Although Run's nutmeg plantations had been completely destroyed and the island was no longer of any economic importance, the British were to keep their flag flying over it until they eventually acknowledged Dutch rights in 1667 (the same year they evacuated Ujung Pandang). By the Treaty of Breda, the British acquired in exchange for Run the island of Manhatten in New Amsterdam, the New York of modern times.

During the first half of the seventeenth century, the VOC controlled the production of spices to ensure the highest possible return for the Company; they decided where spices were to be grown and in what quantity. Overproduction and smuggling were prevented by destroying trees elsewhere. Nutmeg and mace were cultivated only on the Banda Islands, where the number of trees was restricted. Since Ambon's clove production was felt to meet Company requirements (the island was rich in cloves by the sixteenth century), trees were pulled up in the northern islands. This was particularly cruel as it was a Moluccan tradition to plant a clove tree for the birth of a child. Conversely, destroying a tree was believed to foretell doom for the child. On Tidore and Ternate, the original home of the clove, production had been the sole source of income for the islanders. The penalty for cultivating extra spices was death.

There were violent native revolts in response to these

measures, and the Sultan of Ternate was arrested, sent to Batavia, and only reinstated as sultan when he agreed to Dutch terms. This he did in 1652 when he was forced to sign an agreement with the Dutch giving them the right to cut down as many clove trees and food-productive sago palms as they thought necessary.

Further punishment to the islands involved the destruction of sago palms on the Howamohel Peninsula on Seram, which had supplied other islands in the Moluccas with this essential food, and was one of the reasons Seram was known as the 'mother island'. The area became uninhabitable and the population was forced to move to Hitu on Ambon by the Governor of the Moluccas, Arnold de Vlamingh van Oudshoorn, who told islanders that their heads would be put before their feet if they resisted the relocation. De Vlamingh's name became a curse, and to this day Seram is thought to be populated by ghosts.

Adherence to Dutch policies on planting were enforced by the dreaded 'hongi' raids. Sometimes as many as 100 large Bandanese war canoes, the *kora-kora*, set out from the Dutch headquarters of Ambon (Colour Plate 10). These were crewed by local villagers who were rewarded by their Dutch overlords with extra land for supplying men for the raids. Under the command of Dutch officers, they patrolled the islands, uprooting unlicensed trees, and razing the homes and destroying the property of offenders. This restrictive policy on the production of spices was to be enforced until 1824.

Throughout the Moluccas, it was obligatory to sell spices only to the Company at a fixed sum, a sum so low that the islanders were unable to buy the highly priced rice, sago, and cloth imported for them by the Dutch. Smuggling and illegal trading were strictly forbidden, nor was time allowed for attempts at subsistence farming or fishing; it was to be for compulsory spice cultivation.

The result of this rigid Dutch control on spice production and trade was depopulation and penury on once prosperous islands as the people's source of livelihood disappeared. Destruction of local shipping and coastal poverty also led to a big increase in piracy, which by the eighteenth century was strong enough to openly challenge the Dutch. The Moluccas were becoming a wilderness, and with few people to tend them, clove and nutmeg plantations went unharvested and uncared for.

To continue the cultivation of nutmeg and mace, it was necessary to restock decimated Banda and an entirely immigrant population was thus introduced, comprising slave labour from Java and Irian Jaya. Ambonese and Dutch—either VOC employees already in the Indies or recruited from Holland—became the new estate managers. Even the tough Coen was said to be rather dismayed by the calibre of some of the conscripts: out-of-work ruffians, even criminals, in fact anyone who could be coerced into a life which, at one time, spelled almost certain death in the Indies. The Dutch lost their lives just as the Portuguese had in the sixteenth century, by shipwreck, war, and sickness. The granite gravestones paving the floor of Banda Neira's Dutch church, and filling the cemetery, bear testimony to this.

Land on the Banda Islands of Lontar, Ai, and Neira was divided up into plots which were called 'perks'. These were managed by Dutch managers—known as 'Perkeniers'—who worked the land with the help of slave labour. Neither party had any experience in the cultivation of spices and few Moluccans who could advise them were left. The VOC paid a fixed price according to the quality of the spice (this was decided by the Company) to the Perkeniers. Profits to the Company were huge, often exceeding 1,000 per cent, and Perkeniers told of a single nutmeg being worth one silver Dutch crown.

The Moluccans themselves were reduced to a subsistence existence. Their sole source of commerce had been seized by the Dutch. Even the Heeren XVII were to feel that Coen had been 'too energetic' in his Moluccan policies. It was realized that if spice production—and its profits—were not to cease altogether, then sufficient food and cloth at affordable prices had to be provided for the islanders. Rules were slightly relaxed, some interisland trading was re-established, and food crops were allowed to be grown.

In 1658, the Sultan of Palembang in south-east Sumatra was forced to allow the construction of a Dutch fort and the exclusive rights to purchase his pepper. In 1663, the other pepper centre, strongly Muslim Aceh in North Sumatra (and a scourge to the Dutch as she had been to the Portuguese) lost control of pepper exports but managed to retain her independence. In 1663, the Spanish evacuated Tidore. But it took two major expeditions before the Sultan of Macassar (Ujung Pandang) finally accepted Dutch rule in November 1667, agreeing also to the expulsion of the Portuguese and British. Portuguese Malacca fell to the Dutch in 1641 after a six-year blockade of the Strait. Banten signed a treaty in 1684 whereby she, too, agreed to expel all non-Dutch Europeans and to grant total pepper trading rights to the VOC. The British were forced now to leave the pepper town of Banten, as they had left Macassar, and move to Bengkulu on Sumatra's west coast where they were to remain until 1824.

Pepper was the staple item of commerce in the Indies trade of the seventeenth century, with Holland providing around 80 per cent of Europe's needs. At the beginning of the eighteenth century, the Dutch were complete masters of the Indies. They were able to fix clove and nutmeg prices in European and Asian markets, and their annual fleets to Europe were laden with spice. In the later part of the century, 'mountains' of cinnamon and nutmeg were burnt in Amsterdam to maintain prices.

Profits for the VOC fell when European demand for spices declined as the eighteenth century progressed. Fresh meat was becoming available all year round, and dietary habits were changing. As Portugal had found in the previous century, the costs of maintaining a commercial empire with its fleets, forts, and factories, and enforcing a monopoly on spice trading, were huge. Dutch goods traded in the Indies were unwanted and too expensive. Local needs were met by smuggling, and, in spite of heavy penalties, Company employees continued to engage in private trading to augment their below subsistence salaries. Having borrowed heavily for years to continue to pay out dividends, when its charter came to an end in 1799, the VOC had accumulated debts of millions of guilders. These were assumed by the Crown, who took over the administration of what was to be known as the Dutch East Indies. Though the British East India Company (far poorer than the VOC at its inception) had managed only a small part of the trade in Eastern spices, it paid out higher dividends than the Dutch Company.

The East Indies monopoly on the production of spices was broken in 1770 when Frenchman Pierre Poivre, Governor of Mauritius, smuggled out clove and nutmeg seedlings. From Mauritius the clove was taken to Zanzibar by the Arab Harameli bin Saleh, in 1818. This led, in modern times, to Zanzibar and neighbouring Pemba on the East African coast becoming the world's main suppliers of cloves (Indonesia remains the largest producer). Spice seedlings were also acquired for the British in the nineteenth century during their occupation of the Indies and taken to India and Europe. These ventures were successful enough to keep prices at a reasonable level when European demand for spices grew again, and to restrict Dutch profits.

During the Napoleonic Wars, Britain took over Holland's possessions in the Indies. The British occupation was organized from Malacca in 1811 by the Governor-General of India, Lord Minto, and was led by the young officer Thomas Stamford Raffles who was to become Lieutenant-Governor of Java and its dependencies (Plate 17).

While appreciating the enormous forest and mineral resources of the archipelago—its gold, spices, and camphor—and the potential for British trade, Raffles did attempt to improve the welfare of the individual during his years on Java, following Minto's advice: 'While we are in Java, let us do all the good we can' (Hall, 1964). He implemented administrative and legal reforms similar to those used by the British in India, his most important contribution being the introduction of a land tax system which was retained by the Dutch when the Netherlands Indies were returned to Holland after Napoleon's defeat in 1816. This was to be a disappointment for Raffles, who had dreamed of substituting British rule for Dutch in the East.

Leaving his post on Java, Sir Thomas Stamford Raffles (he had been knighted in 1817) was appointed, in 1818, to Bengkulu in south-west Sumatra. This flat, humid, fever-ridden coast was where the English East India Company had planted a settlement

17. Bust of Raffles, from Mildred Archer and John Bastin, *The Raffles Drawings in the India Office Library, London*, Kuala Lumpur, Oxford University Press, 1978.

in 1682 when they were expelled by the Dutch from Banten. It had been thought that Bengkulu would prove to be strategically situated on the China route, but the best way to China was found to be through the Strait of Malacca and not the Sunda Strait. The province's agricultural potential was never realized either, though Raffles did his best to create a model colony. He attempted to raise the standard of living, agriculture was expanded and encouraged by the English Company, spices as well as coffee and sugar were grown, and planters could export their produce free of duty to Britain.

Upon becoming Lieutenant-Governor of this insignificant province, Raffles wrote to his friend William Marsden (whose *History of Sumatra* was to become the definitive work on the island) that Bengkulu 'is without exception the most wretched place I ever beheld'. Three of Raffles's four children were to die here during his term of office.

Spice cultivation had actually been started in Bengkulu province by the Englishman Edward Coles in 1798. He had planted, with some success, imported nutmeg and clove seedlings on his estate at Pematang Balam. This was in a much healthier location, 12 miles out of Bengkulu, and Raffles was to later acquire it for the East India Company. By 1821, Raffles was more optimistic of Bengkulu, writing again to Marsden of his expanding agricultural, spice, and coffee policies, his cattle, and the good roads that gave him access from the British Fort Marlborough to his house. Called 'The Abode of Peace', he compared it to living in the country, 'cool and retired'. His second wife Sophia was to describe it as being 'surrounded by [spice] plantations. The clove trees as an avenue to a residence are perhaps unrivalled, their noble height, the beauty of their form, the spicy fragrance with which they perfume the air, produce, in driving through a long line of them, a degree of exquisite pleasure' (Colour Plate 11).

By the Anglo-Dutch treaty of 1824, the Netherlands ceded to Britain all her factories in India. Holland's possessions in the Indies were returned to her, including Bengkulu. As Dutch interest was centred on the more fertile and productive Padang, north of Bengkulu, Raffles's estate at Pematang Balam was left to run wild. He himself was to go on to found Singapore for Britain.

With the spice trade in decline, there was little commercial activity in the Moluccas during the British occupation of the Indies. The most significant activity was perhaps the arrival of new immigrants to Banda, who took over trading and slightly revitalized the economy. Robert Farquhar, Deputy Commercial Resident of Banda at the end of the eighteenth century, was to note in reports that a number of 'Tanimbarese, Balinese, Butonese, Buginese ... besides many emancipated slaves ... slid by degrees into the island' (W. G. Miller, 1980). Banda was, however, to continue exporting her nutmeg and mace, and as much as 269 tons of nutmeg and 68 tons of mace left the island in 1854.

Farquhar also noted the ships that came from Batavia in December and January to trade for Banda's spices. They carried a wide range of goods: there was cotton from India in large quantities (that and opium continued to be the main exports from British India to the Moluccas), Chinese porcelain, iron, thread, soap, beef, butter, French and Spanish wine, salt, pepper, rice, wheat, and sugar. After their trade with Banda, these ships would sail on to other islands in the Moluccan archipelago, and return with sea slugs, birds' nests, pearls, and Bird of Paradise plumes. They would then sail back to Batavia in early June to catch the Chinese trade before the junks departed for home on the monsoon.

An American naturalist, Henry Ogg Forbes, who visited Banda in 1881 talked of 'its atmosphere ... charged with aromatic exhalations', and of the vast white plantation houses of the Perkeniers where mace was peeled and dried in the sun to make 'as delightful a cargo as could be wished' (quoted in Hanna, 1991). In modern Banda, some of these old mansions still stand, graceful ruins now, but a reminder of their colonial past.

7
Spices in Culture

THE spices that grew so profusely in Indonesia, with their per-
fuming, flavouring, healing and colouring properties, were—and
still are—widely employed in the islands in ceremonies of rites of
passage, as temple offerings, in the preparation of traditional
medicines, for the dyeing of ceremonial textiles, and as design
elements in many forms of arts and crafts.

Jamu

Spices have been used as traditional medicines in the East for
thousands of years. Camphor from Sumatra was traded as a drug
by the Arabs, and it was still being exported as a medicinal spice
during later Dutch rule, as were the tree resin benzoin and the
yellow rhizome of turmeric. Several Indonesian spices are today
on the World Health Organization's list of the most used medicinal plants in the world, and are legally prescribed and included
in the drugs of many countries.

Jamu is the Indonesian term for a traditional herbal medicine
that provides health and beauty care for a large section of the
population. The practice of preparing and taking *jamu* originated
in the seventeenth-century courts of Central Java, Indonesia's
cultural heartland, where ladies of the royal palaces blended
spices, herbs, and plants into formulas that would maintain their
health and beauty. These formulas were passed down and, over
the years, became increasingly accessible to the common people.
Today, almost 80 per cent of the population use some form of
jamu. The popularity of *jamu*, which is exported to other South-
East Asian countries, as well as to Australia and Europe, reflects a
modern tendency towards a more natural form of health care
long in use in the East.

During the Dutch occupation of the eighteenth century, the

curative powers of Indonesian plants and spices were documented by the German botanist Georg Everhard Rumphius while in the service of the Dutch East India Company in the Moluccas. He relied, to a certain extent—as do Indonesia's respected local *dukun*, or medicine men—on the dictum that plants are marked by God or nature with a symbol indicating how they should be used (Beekman, 1981). Hence, a moss growing on a windy site is prescribed for flatulence, a shrub producing a red dye is thought to stop bleeding, and the yellow rhizome of turmeric is used in the treatment of the yellowing disease jaundice.

The plants needed for the more common *jamu* grow in every garden, and simple herbal formulas for common ailments have been made at home for centuries. Leaves are picked and roots dug up, ground or pounded, then infused to make up recipes learnt from grandmothers and mothers. Of the spices employed, turmeric is known as the champion of *jamu*. There is an endless list of products in which it is used, including preparations to relieve angina, soothe ulcers and colic, and cure liver problems, anaemia, eczema, and dysentery. Because of its warming and soothing properties, ginger has widespread therapeutic qualities: as an aid to digestion, for the relief of headaches, as a liniment for rheumatism, as a remedy for colic and coughs, as a tonic, and as an antidote for snake bites. Cardamon makes a refreshing gargle, while cloves provide flavouring and disguise unpleasant tastes. Rumphius documented the supposedly hallucinogenic and aphrodisiac properties of the nutmeg, whose lacy covering of mace is a sedative and relieves insomnia. Ancient Chinese prescriptions still employ the fragrant wood of cinnamon.

Jamu is produced by small family-run firms, who compete with the mass-marketing of the large, modern factories. Both large and small operations continue to rely on recipes passed down through generations, though all take pride in experimenting with new blends and products. The best companies grow their own plants so that they can supervise the picking and washing that are done before drying and pulverizing. The blending of the ingredients into the various formulas is the final process. In the commercial production of *jamu*, some 200 species of selected plants are used: spices, flowers, roots, grasses, and seeds, and, in the larger *jamu* factories, as much as 700 tonnes of selected plant ingredients are processed monthly. These go into making hundreds of different preparations, for *jamu* is employed for every malady as well as for preserving good looks and restoring vigour. Products are applied externally as cosmetics for the face, hair, and body or taken orally in the form of capsules, powders, or infusions.

Jamu is mass-marketed by large companies and widely sold in supermarkets and drug stores. It can also be bought from *warung*—tiny roadside stalls and eating places—or delivered to the door by the *jamu gendong* (Plate 18). This lady, attired in a traditional sarong, is to be found throughout Java selling *jamu*. She carries on her back a large basket filled with bottles in which she has mixed a selection of the most popular ready-to-drink preparations for common ailments and complaints. These she sells by the dose.

Kretek

The manufacture of a distinctly Indonesian cigarette, the *kretek*, in which cloves are mixed with tobacco, began in Indonesia early in the nineteenth century as a small cottage industry. Today, this industry is a huge concern producing around 18 billion *kretek* annually. Several large, automated companies employing thousands of workers, many of whom are women, have mostly replaced the small cottage manufactories.

Because the *kretek* market is largely domestic—the warm aroma of burning cloves is pervasive throughout the country—about 80 per cent of Indonesia's annual output of around 150 000 tonnes of cloves is utilized in the manufacture of *kretek*.

To make this unique cigarette, the cloves are first steamed to soften them, then dried till they achieve the correct moisture content. Next, the cloves and tobacco are shredded and combined to make a blend that is 60 per cent tobacco, 40 per cent clove. A final top-dressing of spices is sprayed on to enhance the flavour. Rolling and packaging remain labour-intensive tasks,

18. A *jamu gendong*, Jakarta. (Ganesha Volunteers.)

with a good worker rolling as many as 8,000 sticks a day (Colour Plate 12).

Clove Crafts

The making of models with cloves begun as a cottage craft in Ambon in the eighteenth century, and is a skill passed down through families. Dried cloves—thousands of them—are strung tightly together on wires which become completely invisible. These models were originally copies of the Bandanese *kora-kora* canoes, which were used in local trading or by the Dutch for punitive raids. The range of models has expanded over the years to include houses, birds, flowers, baskets, and large sailing vessels that, in perfect detail, often bear oars, flags, lanterns and tiny figures, all made out of cloves (Colour Plate 13).

Traditional Colouring

The colour yellow has early associations with kingship in South-East Asia. It is the sacred colour of the Hindus, and the traditional colour of royalty in Northern Thailand, Laos, and Malaysia. In coastal Malay sultanates, yellow was restricted to royal use, and at one time could only be worn by princely families in Indonesia's islands of Sumatra, Kalimantan, and Sulawesi.

Yellow rice, coloured and flavoured with the rhizome turmeric, is prepared for ceremonial occasions and presented as a slender yellow cone. The top of this is sliced off and given to the most important guest (Colour Plate 14). Turmeric also colours the yellow rice used in temple offerings on the Hindu island of Bali, where cyclic festivals and ritually prepared offerings play a large part in the lives of the people. Aragon and Taylor (1991) talk of the *perahu lancang kuning*, or ceremonial yellow boats, that were displayed at harvesting ceremonies in Sumatra.

For hundreds of years, turmeric has also been used in Indonesia to dye traditionally woven *ikat*, in which bound yarn—*ikat* literally means to bind, or tie—is dyed before being woven into cloth. In the royal courts, ceremonial clothes made from rich silk were also dyed in turmeric.

Turmeric produces a reddish-yellow colour (Colour Plate 15). The powdered rhizome, soluble in water, is used in solution and applied by vigorous kneading. An acid such as lime or tamarind acts as a mordant to help fix the dye, which is very fugitive. Dyeing was at one time a craft practised only by men, who jealously guarded the process as they mixed, touched, tasted, and smelt their potions.

Textile Design

Indonesia has long produced fine traditional textiles. The design register for her woven and batik cloth, as with her silver and wood-carving, owes much to the country's abundant plant and animal life, a stylistic affinity shared with the Malay Peninsula with whom she has been so long associated. It has also been influenced by the Islamic faith which proscribes the use of the human or animal form as a decorative motif.

A popular vegetal motif is the flowering tree, with leaves, flowers, and intricate tendril ornamentation, and there was certainly design influence, too, from trading contacts with India, China, and later Europe.

A traditional and common woven floral design is the *cengkeh* (clove) flower. Selvanayagam (1990) describes it as looking like a tiny cross made up of four or five squares, usually woven in gold thread. Too small to be used alone, the *cengkeh* motif fills empty spaces in the pattern or is put into the centre of another, larger motif (Colour Plate 16).

The batik process of wax resist is a Javanese craft. Cotton or silk is patterned through successive dyeing, with areas to be left undyed covered with a protective wax coating. The nutmeg motif on the splendid north coast Javanese batik in Plate 19 is distinctively lifelike. The fruit on its sinewy branches has split open to reveal the seed, while fluttering between the fruits are Banda's nutmeg pigeons, depicted here as unrealistically small.

19. Nutmeg tree motif on a batik sarong from the north coast of Java, mid-nineteenth century. (Tropenmuseum, Amsterdam.)

Silver

The island of Java is mentioned as a source of silver in Chinese annals of the Tang Dynasty (AD 618–906), and Sumatra has long had rich reserves. Techniques in metal-working came with early Indian trade, gold being worked in the archipelago long before silver which was used initially to decorate the more precious metal.

Silver bars were currency in the seventh-century Sumatran kingdom of Srivijaya. During later colonial times, silver coins carried to the Indies by the English East India Company were often melted down for silverwork.

There were design influences from China and from Arabia (with the spread of Islam from that area) and, as with cloth, a similarity to Malaysia in decorative motifs used on Indonesian silver. Plant and vegetal forms were widely utilized, often in a scroll-like form, with the clove head a commonly featured motif (Plate 20).

Silver was also made into jewellery and court regalia. The scabbard of the sacred Javanese kris, or dagger, was often encased in silver richly embossed with a floral pattern. Among five basic decorative patterns for the kris blade, which had attributed to it a mystical quality, were nutmeg and ginger flowers.

Silver was widely utilized in the production of the accoutrements for the ceremony of betel-chewing. Crafting the variously

20. Detail of a rectangular silver pillow end decorated with scrolls and fruits resembling clove heads. (National Museum, Singapore.)

shaped utensils used to hold the ingredients provided enormous scope for the decorative skills of the South-East Asian silversmith.

Betel Chewing

The betel quid has been chewed and ceremonially exchanged throughout South-East Asia for centuries: there are mentions of it in Indian manuscripts dating to 500 BC, and in eighth-century China guests were offered betel rather than tea (Jessup, 1990). *Pinang*, the large nut of the areca palm (*Areca catechu*) (Colour Plate 17), is finely sliced and wrapped in the aromatic leaf of the betel vine (*Piper betle*). Crushed lime is added, probably obtained originally from burnt coral or shells, and sometimes gambier, which is an extract from the leaves of the *Unicaria gambir* shrub. To this basic quid, tobacco and spices may be added for flavour. The spices used can vary, though cinnamon, cardamom, slivers of nutmeg, or slices of turmeric are the most popular. 'Cardamoms and cloves compose part of the articles in the sirih box of a person of condition,' wrote Raffles (1817). *Sirih* is the Indonesian term for the quid. Chewed over a long period, it has a mildly stimulating effect. Habitual use leads to staining of the mouth and teeth.

Betel chewing and exchange accompanied rituals of circumcision, death, and burial throughout the islands and the toothfiling rite of Hindu Bali, where it was also offered to the spirits. It was an important part of marriage arrangements, when betel might be exchanged as an early sign of interest between a boy and girl at the start of a courtship, then offered by both families on betrothal, and with more ceremony, to family and guests at the wedding itself. By the fifteenth century, the *pinang* ceremony had become an indispensable part of court functions. In nineteenth-century Ternate, whenever the Sultan visited the Dutch Resident, he was offered betel from a silver service.

Special containers and implements were made for all the ingredients of betel chewing: triangular-shaped receptacles for holding the betel leaf, little boxes for the tobacco, lime, and spices,

21. A nineteenth-century Javanese silver *sirih* set, from the collection of Mr and Mrs Joern-Uwe Neiser. (Kirsten Nieser-Schlaudraff.)

sometimes in the shape of fruits (see the boxes in Plate 21), nutcracker-like cutters for slicing the areca-nut and gambier, and spittoons for depositing the saliva which flowed freely during chewing. *Sirih* sets reflected status. At the most ordinary level, a labourer would tie his betel chew into a corner of his handkerchief (Raffles, 1817), while wooden boxes or palm leaf baskets formed simple everyday containers, as did those woven from rattan and grasses. They were often made of brass for household use. In the courts of the archipelago, ornate sets of silver and gold embossed with jewels were used, and sometimes given as gifts of great value to foreign rulers and visiting dignitaries (Jessup, 1990).

In modern times, the trade in the spices of Indonesia remains an active one. Spices and their essential oils are profitably exported all over the world, with America and Europe as large consumer countries. Pepper remains the most important in terms of both volume and value. The spice entrepôt for the region is now Singapore, at the tip of the Malay Peninsula, very near to the port of Malacca that once controlled the Strait.

Select Bibliography

Aragon, Lorraine V. and Taylor, Paul Michael, *Beyond the Java Sea: Art of Indonesia's Outer Islands*, Washington, DC, The National Museum of Natural History, Smithsonian Institution, 1991.

Bastin, John and Archer, Mildred, *The Raffles Drawings in the India Office Library, London*, Kuala Lumpur, Oxford University Press, 1978.

Beekman, E. M., *The Poison Tree: Selected Writings of Rumphius on the Natural History of the Indies*, Amherst, University of Massachusetts Press, 1981; reprinted Kuala Lumpur, Oxford University Press, 1993.

Boxer, C. R., *The Portuguese Seaborne Empire 1415–1825*, London, Hutchinson and Co. Ltd., 1969.

Brissenden, Rosemary, *Patterns of Trade and Maritime Society Before the Coming of the Europeans: Studies in Indonesian History*, Victoria, Pitman, 1976.

Burkill, I. H., *A Dictionary of the Economic Products of the Malay Peninsula*, Singapore, Governments of Malaya and Singapore, 1966.

Chittick, H. N., *Early Ports in the Horn of Africa*, Kenya, The British Institute in Eastern Africa, 1979.

———, *East African Trade with the Orient*, in D. S. Richards, *Islam and the Trade of Asia*, University of Pennsylvania Press, 1970.

Cortesão, Armando (trans.), *The Suma Oriental of Tomé Pires: An Account of the East, from the Red Sea to Japan, written in Malacca and India in 1512–1515*, London, Hakluyt Society, 1944.

David, Elizabeth, *Spices, Salt and Aromatics in the English Kitchen*, London, Penguin Books, 1970.

Eiseman, Fred B., *Bali: Sekala and Niskala*, Vol. I, Singapore, Periplus Editions, 1988.

Ellen, Roy F., *Sago Subsistence and the Trade in Spices: A Provisional Model of Ecological Succession and Imbalance in Moluccan History*, London, Havers Press, 1979.

———, 'The Trade in Spices', *Indonesia Circle*, March 1977.

Galvão, Antonio, *A Treatise on the Moluccas (c.1544)*, probably the preliminary version of his lost *História Das Molucas*, Volume III, Rome, Jesuit Historical Institute, 1970.

Glanville, Philippa, *Silver in Tudor and Early Stuart England*, London,

Victoria and Albert Museum, 1990.

Hall, D. G. E., *A History of South-East Asia*, New York, Macmillan and Co. Ltd., 1964.

Hanna, Willard A., *Indonesian Banda, Colonialism and Its Aftermath in the Nutmeg Islands*, Banda Naira, Yayasan Warisan dan Budaya Banda Naira, 1991.

Jessup, Helen Ibbitson, *Court Arts of Indonesia*, New York, The Asia Society Galleries, 1990.

Jourdain, John, *The Journal of John Jourdain, 1608–1617 describing his experiences in Arabia, India and the Malay Archipelago*, edited by William Foster, Cambridge, Hakluyt Society, 1905.

Linschoten, Jan Huygan van, *Navigatio ac Itinerarium in Orientalem sine Lusitanorum*, Amsterdam, 1592.

McBain, Audrey Y., 'Oriental Artifacts Discovered in Madagascar', *Arts of Asia*, September–October 1988.

McKay, Elaine, *Sixteenth to Eighteenth Centuries: The Significance of the Coming of the Europeans: Studies in Indonesian History*, Victoria, Pitman, 1976.

Marsden, William, *The History of Sumatra*, London, 1738; reprinted Kuala Lumpur, Oxford University Press, 1966, and Singapore, Oxford University Press, 1986.

Maxwell, Robyn, *Textiles of Southeast Asia*, Canberra, Australian National Gallery and Melbourne, Oxford University Press, 1990.

Miller, J. Innes, *The Spice Trade of the Roman Empire 29 BC to AD 641*, London, Oxford University Press, 1969.

Miller, W. G., 'An Account of Trade Patterns in the Banda Sea in 1797', from an unpublished manuscript in the India Office Library, *Indonesia Circle*, November 1980.

Past Worlds: The Times Atlas of Archaeology, Verona, Times Books Ltd.,1988.

Purseglove, J. W.; Brown, E. G.; Green, C. L.; and Robbins, S. R. J., *Spices*, Vols. I and II, New York, Longman, 1981.

Raffles, Thomas Stamford, *The History of Java*, London, 1817; reprinted Kuala Lumpur, Oxford University Press, 1965 and 1978, and Singapore, Oxford University Press, 1988.

Schoff, W. H. (trans.), *The Periplus of the Erythraean Sea*, London, 1912.

Rosengarten, Frederic Jr., *The Book of Spices*, Wynnewood, Livingstone Publishing Company, 1969.

Selvanayagam, Grace Inpam, *Songket: Malaysia's Woven Treasure*,

Singapore, Oxford University Press, 1990.

Suparto, Darmi, *Jamu in the Health and Beauty Care of the Javanese Woman*, Jakarta, privately published, 1974.

Van Leur, J. C., *Indonesian Trade and Society*, The Hague, W. van Hoeve Ltd., 1955.

Veevers-Carter, W., *Riches of the Rain Forest*, Singapore, Oxford University Press, 1984.

Wallace, A. R., *The Malay Archipelago*, London, Macmillan & Co., 1869; reprinted Singapore, Oxford University Press, 1986.

Index

SULAWESI (CELEBES)

HALMAHER (JAILOLO)

TERNATE
TIDORE
MOTI
MAKIAN

MOLUCCA SEA

BACAN

OBI

SULA ISLANDS

BURU

AMBON
ISLAND   Am

BANDA SEA

THE MOLUCCAS